International Research Advancement Network 2019

AA006643

IOP Conference Series: Materials Science and Engineering
Volume 607

Online Virtual Conference
15 May 2019

Editors:

Reazul Haq Abdul Haq
Ahmad Mubarak Tajul Arifin
Mohd Nasrull Abdol Rahman
Mohd Fahrul Hassan
Ishkrizat Taib

ISBN: 978-1-5108-9484-6
ISSN: 1757-8981

Printed from e-media with permission by:

Curran Associates, Inc.
57 Morehouse Lane
Red Hook, NY 12571

Institute of Physics
Dirac House, Temple Back
Bristol BS1 6BE UK

Phone: 44 1 17 929 7481
Fax: 44 1 17 920 0979

techtracking@iop.org

Additional copies of this publication are available from:

Curran Associates, Inc.
57 Morehouse Lane
Red Hook, NY 12571 USA
Phone: 845-758-0400
Fax: 845-758-2633
Email: curran@proceedings.com
Web: www.proceedings.com

Table of contents

Volume 607

International Research Advancement Network 2019

15 May 2019, Online Virtual Conference

Accepted papers received: 2 August 2019
Published online: 30 August 2019

Preface

Papers

Material Science and Engineering

IOP Conf. Series: Materials Science and Engineering **607** (2019) 011001 doi:10.1088/1757-899X/607/1/011001

PREFACE

The 1st International Research Advancement (IRA) Network 2019 is organized by Edra Resources located in Johor, Malaysia. This conference is an example of a budget conference with no costs for flying around the world, accommodations or getting a visa as it conducted via virtual. Although this conference will not provide the same experience as an in-person conference, but this still leads to a lively interaction between the participants. This conference were held on 15th May 2019 in virtual presentation as the main purpose of the IRA Network 2019 is to bring together researchers, academia, engineers and natural scientists from all over the world, interested in the field of Science, Technology and Engineering.

There are 25 papers from various topics including material science, mechanical and automotive engineering, Manufacturing Technology, Process and Systems Advanced Machining and Rapid Prototyping, Industrial Engineering and Operations Management and etc. Only 12 papers are selected to go through a strict peer reviewed process, and these papers will be published in conference proceedings. We would like to thank the organizing committee and the members of reviewers for their kind assistance in reviewing the papers.

The Editors

Dr. Reazul Haq Abdul Haq
Dr. Ahmad Mubarak Tajul Arifin
Dr. Mohd Nasrull Abdol Rahman
Dr. Mohd Fahrul Hassan
Dr. Ishkrizat Taib

IOP Conf. Series: Materials Science and Engineering **607** (2019) 011002 doi:10.1088/1757-899X/607/1/011002

Peer review statement

All papers published in this volume of *IOP Conference Series: Materials Science and Engineering* have been peer reviewed through processes administered by the proceedings Editors. Reviews were conducted by expert referees to the professional and scientific standards expected of a proceedings journal published by IOP Publishing.

3D Printer parameters analysis for PCL/PLA filament wire using Design of Experiment (DOE)

Reazul Haq Abdul Haq [1,2], Omar Mohd Faizan Marwah [1,2], Mohd Nasrull Abdol Rahman [1], Ho Fu Haw [1], Haslina Abdullah [1], Said Ahmad [1]

[1] Department of Manufacturing and Industrial Engineering, Faculty of Mechanical and Manufacturing Engineering, Universiti Tun Hussein Onn Malaysia, 86400 Parit Raja, Batu Pahat, Johor, Malaysia
[2]Additive Manufacturing Research Group (ADDMAN). Department of Manufacturing and Industrial Engineering, Faculty of Mechanical and Manufacturing Engineering,, Universiti Tun Hussein Onn Malaysia, 86400 Parit Raja, Batu Pahat, Johor, Malaysia

*Corresponding author email: reazul@uthm.edu.my

Abstract: Fabrication of medical implant using composite Polycaprolactone (PCL)/ Polylactic Acid (PLA) through a process of fused filament fabrication (FFF) is the future of medical field. In this study the capabilities of FFF process and optimization of the Parameter was investigated to achieve the 20mm x 20mm x 10mm dimensional accuracy of printed composite material. Design of experiment using technique full factorial with two level design is used to study the interaction of factor which is print temperature (A), height of deposition (B), print speed (C) and travel speed (D). By using DOE, the total 20 run include with 4 center point was investigated by extruding the composite material using 3D printer. The dimension accuracy is tested using flexible measuring machine. From the sample, the ANOVA analysis of variance is divided into 3 responses, length, width and height. Factor model A, C and BD show the significant effect on length and for width response, factor model effected the dimensional accuracy is A, C, AC, and AD. Meanwhile for height respond only factor model A and C show the significant effect on the respond. Based on the analysis, one optimum parameter run is selected among ten of the suggested run. The optimum parameter value consists of 136°C print temperature, 0.12mm deposition of height, 20mm/s printing speed and 22mm/s travel speed. The confirmation run analysis this optimum parameter shows the precise dimension with average margin error below that 15% for length, width and height respond.

Keywords: Fused filament fabrication (FFF); Design of experiment (DOE); PCL; PLA; FDM

1. Introduction

In the past few years, 3D printing technology is growing rapidly. The ability of this technology allows fabrication of custom-made implant and medical devices, suddenly realize the world about the potential of this technology for biomedical field. Additive manufacturing had becoming a new technology that provides crucial impact on biomedical implant. This technology gives an opportunity to customized implant according to the specific patient [1][5].

Before additive manufacturing had been introduced to the biomedical implant fabrication, the previous method was using the conventional process. The conventional process is too complicated due to much fabrication stage but only for a single implant. This typical process involves several steps such as taking images for the tissues, produce mold, calculating for the tissue geometry and even machining using Computer numerical control machine. This conventional process is time consuming and increased the price tag for the implant [2].

Since the introduction of additive manufacturing for fabricating biomedical implant, the process become easier, low cost, less time consuming and it enable directly customized implant for the patient. However, this immature technology has substantial challenges in term of type of material and material printed capability using 3D printer. To solve the problem regarding this technology, several studies need to be done in term of new type of material composite and printed capability for the new material using 3D printer. Currently a composite material called

Polycaprolactone (PCL) / Polylactic acid (PLA) is a choice for the implant material.

In medical implant fabrication, the material must be biocompatible and capable to undergoing reaction under physiological condition of body. From the properties of material, Polycaprolactone (PCL) / Polylactic Acid (PLA) composite have the characteristic that suitable for human implant [3]. Despite this composite excellent in term of characteristic, the major problem is the optimization process parameter using 3D printer. The parameter setting such as, print temperature and feed rate have direct influence on the printed capabilities for this composite.

Beside optimization process parameter, the main issue regarding this composite is the printed accuracy for producing implant. Every medical device such as implant needs accuracy in dimension. The accuracy of dimension related to the precision in term of extrusion of the material and parameter setting [4][6]. This work focuses on study in optimization parameter of PCL/PLA material in term of print temperature, deposition height, print speed and travel speed by using 3D printer. The dimensional accuracy was analyzed based on replication sample 20mm x 20mm x 10mm.

2.0 Methodology

2.1 Material Fabrication

Material fabrication is the procedure for preparation the composite material PCL/PLA before the fused filament fabrication process begins by using extrusion process. The processing material is based on extruding the filament using single screw extruder, cooling it using water bath and pull it using pulling roller to get the constant size of filament. The objective of this process is to produce the material filament in range of 1.75 mm to 1.8 mm diameter. Figure 1 below shows the picture of machine use for fabrication of material.

Figure 1: Material fabrication setup

2.2 Designing and printing sample

Designing the sample is first step in fused filament fabrication process. For designing the sample CAD software is needed to drawn 2 x 2 x 1 cm (20 x 20 x 10 mm) cube. Using this software, the CAD model are converted to STL file before the file converted to G-Code file. Figure 2(a) shows the CAD model draw using Solidwork 2014 with dimension unit mm. Next, the test sample is fabricated using FlashForge creator pro 2016 machine. Figure 2(b) shows the 3D printer use in this study.

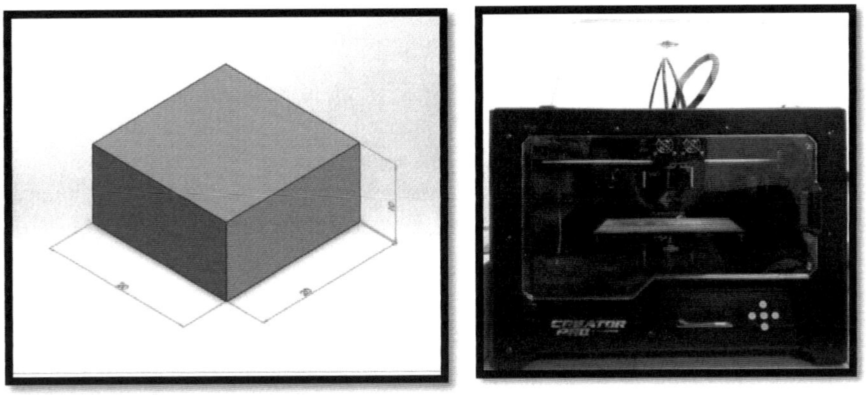

Figure 2: (a) Sample cube 20mm x 20mm x 10mm drawn using Solidwork 2014. (b) FlashForge Creator Pro machine.

2.3 Design of experiment (DOE)

Design expert software version 10 has been use as an aided software to analysis the parameter. In this experiment full factorial with 2 level of factor studies and 4 center point block has been choose to run experiment. By using 4 factor with 3 respond which is length, width and height the total run is 16 run based on 2x2x2x2. However, because 4 center point block, the total run will be 20 run. From the literature review, the range of the factor is tabulated in Table 1.

Table 1: Range of factor parameter

No.	Parameter	Low level	High level
1.	Print temperature (°C)	130	160
2.	Height of deposition (mm)	0.1	0.2
3.	Printing speed (mm/s)	20	50
4.	Travel speed (mm/s)	20	50

IOP Conf. Series: Materials Science and Engineering **607** (2019) 012001 doi:10.1088/1757-899X/607/1/012001

2.4 Dimensional Accuracy

After the printing process using FFF 3D printer. The dimensional accuracy need to be tested in order to find the best variable that produce the most accurate printed model. To test the dimensional accuracy for printed model 2 x 2 x 1 cm cube, Flexible measuring machine Mitutoyo QM353.

3.0 Results and Discussion

3.1 Result on collected respond data

According to the Table 2 below, 4 factor which is print temperature, height of deposition, printing speed and travel speed is represent as factor 1, factor 2, factor 3 and factor 4. While for the respond which is length, width and height is represent as respond 1, respond 2 and respond 3. From the respond data, the data shows run number 3 is the best dimensional accuracy according the cube model dimension 20 mm x 20mm x 10. Data response measured is tabulated as shown in Table 2.

Table 2: Result on the collected data response

No of Run	Factor 1 Print temperature (^{O}C)	Factor 2 Height of deposition (^{O}C)	Factor 3 Printing speed (mm/s)	Factor 4 Travel speed (mm/s)	Response 1 Length (mm)	Response 2 Width (mm)	Response 3 Height (mm)
1	160	0.1	50	20	16.4398	17.9482	10.5260
2	160	0.2	50	50	13.9620	16.8902	10.5804
3	130	0.1	20	20	19.9835	20.0280	10.2245
4	145	0.15	35	35	18.8990	19.3878	10.4840
5	145	0.15	35	35	17.9098	18.5875	10.4525
6	145	0.15	35	35	19.2025	20.0735	10.5732
7	160	0.1	50	50	18.5384	16.8374	10.4906
8	130	0.1	20	50	19.9365	20.0735	10.3735
9	160	0.2	20	50	16.8490	16.7901	10.5582
10	130	0.2	50	20	19.8275	19.5895	10.604
11	130	0.2	20	20	20.1790	20.0775	10.2985
12	160	0.2	20	20	19.5435	19.8735	10.5395
13	160	0.2	50	20	16.3263	18.7967	10.5745
14	130	0.2	50	50	17.8035	20.1085	10.5435
15	145	0.15	35	35	19.1024	20.0845	10.6530
16	160	0.1	20	20	17.6525	19.6920	10.4968
17	130	0.1	50	50	18.8510	20.0410	10.6865
18	130	0.1	50	20	19.3800	19.9385	10.5865
19	160	0.1	20	50	19.6870	19.9880	10.4880
20	130	0.2	20	50	20.0780	20.0460	10.4750

3.2 Surface Morphology

From the experiment, there were certain inconsistencies and error happens due to the incorrect parameters that resulted to a damaged printed part. Figure 3(a) shows the effect of gap between infill and outline. In 3D printing the printed part is created by combining the layer infill and the parameter outline. Typically, in the standard printing process the infill layer is printed faster than the parameter outline. Because of this step of process, the layer infill need to have much time to bond with the parameter outline. From the morphology analysis on 20 printed sample, several sample with higher printed speed is effected on this defect.

A part from that, the effect of stringing or oozing is shown in Figure 3(b). Stringing effect occur because of the extruder temperature is too high when printing part using FFF 3D printer. If the temperature is too high the material become less viscous and will leak from nozzle during the retraction. For this experiment only higher temperature with higher travel speed are effected from this defect.

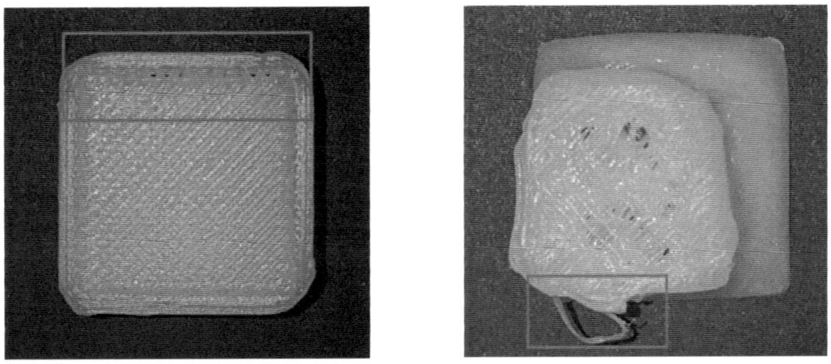

Figure 3: (a) Gap between infill and outline effect; (b) Stringing or oozing effect

Figure 4(a) shows the effect of overheating on the printed sample. Overheating effect occur because of the print temperature is not suitable with the material. When the material extruded from the nozzle, it must achieve the correct balance between temperature of nozzle and the cooling time of the extruded part. However, in this case, the material did not have the right balance of cooling time. This problem happens on run with high temperature for example run number 1 and number 2.

Meanwhile, Figure 4(b) shows the effect of rough corner and warping. Rough corner issue typically occurs because of the material not cooled quickly. Meanwhile for warping issue, the problem always pointed on bed temperature problem. In this experiment, the rough corner were detected on every sample. This defect occurs because of PCL/PLA composite take a lot of time to cool or solidified due to its material properties. For warping issue, the problem occurs because during the experiment the heated bed platform are off due to the low temperature of PCL [7][8].

Figure 4: (a) Overheating Effect; (b) Rough corner and warping effect

3.3 Analysis of Length response

Figure 5 shows the half normal plot graph for length response analysis. From the plotted graph, the significant model effect which are point A, C, BD has been selected for the next evaluation using analysis of variance and Table 3 summarize the data of ANOVA analysis for length response.

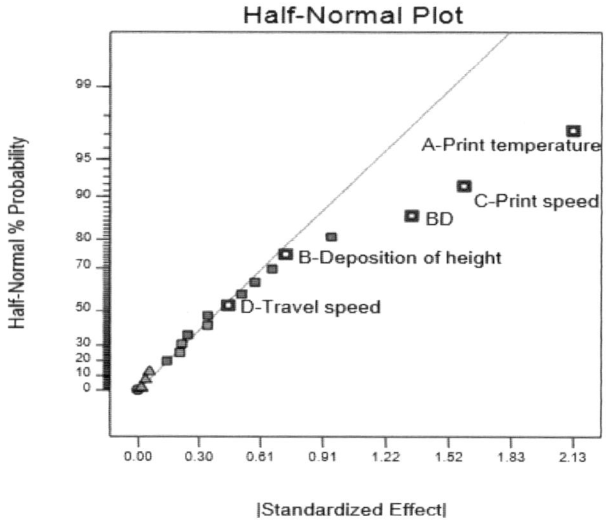

Figure 5: Half normal plot graph for length response

Table 3: ANNOVA analysis for length response

Response : length					
ANOVA for selected factorial model					
Analysis for variance table (Partial sum of square)					
Source	Sum of squares	df	Mean Square	F Value	Prob>F
Model	38.57	5	7.72	9.41	0.0006 *
A-print temperature	18.15	1	18.15	22.14	0.0004
B-Deposition of height	2.18	1	2.18	2.65	0.1272
C-Print Speed	10.21	1	10.21	12.46	0.0037
D-travel speed	0.82	1	0.82	1.00	0.3349
BD	7.21	1	7.21	8.80	0.0109
Curvature	0.37	1	0.37	0.45	0.5152
Residual	10.66	13	0.82		
Lack of Fit	9.60	10	0.96	2.73	0.2208**
Pure error	1.05	3	0.35		
Cor Total	49.59	19			

*significant * non-significant ***

From the ANOVA analysis the model F –value of 9.41 implies the model is significant. Beside that there are only 0.06 % chance that an F-value this large could occur due to noise. In this case model A, C, and BD are significant model term because the values of Prob>F less than 0.0500. However, the other model B and D is insignificant model because the value of Prob>F is large than 0.0500.

For the curvature F value is 0.45 implies the non-significant value. From the table it shows that there are 50% chance a Curvature F>value this large could occur noise. In other analysis part, the "lack of Fit F-value" of 2.73 implies the lack of fit is non-significant relative to the pure error. Chance that a lack Fit F- value this large could occur due to noise is only 22.08%. However, the non – significant lack of fit is still good.

3.4 Analysis of width response

Figure 6 shows the half normal plot graph for width response analysis. From the plotted graph the significant model effect which are point A, C, AC and AD and Table 4 summarize the data of ANOVA analysis for width response.

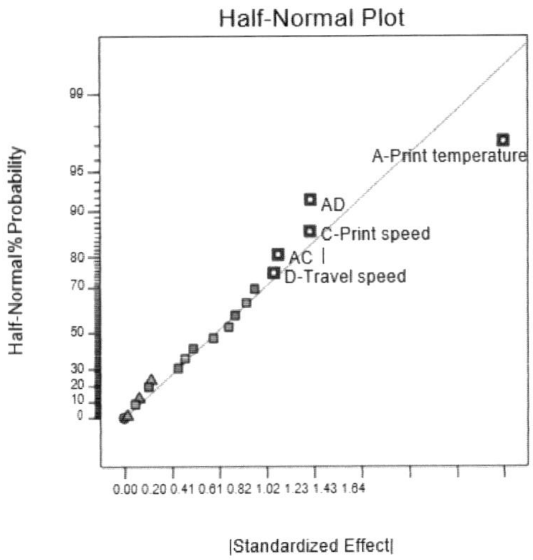

Figure 6: Half normal plot graph for width response

Table 4: ANNOVA analysis for width response

Response : Width					
ANOVA for selected factorial model					
Analysis for variance table (Partial sum of square)					
Source	Sum of squares	df	Mean Square	F Value	Prob>F
Model	19.31	5	3.86	7.05	0.0022*
A-print temperature	10.70	1	10.70	19.54	0.0007
C-Print speed	2.57	1	2.57	2.57	0.0493
D-Travel speed	1.67	1	1.67	3.05	0.1043
AC	1.77	1	1.77	3.23	0.0954
AD	2.59	1	2.59	4.73	0.0486
Curvature	0.42	1	0.42	0.77	0.3957
Residual	7.12	13	0.55		
Lack of Fit	5.61	10	0.56	1.11	0.5244**
Pure error	1.51	3	0.50		
Cor Total	26.85	19			

3.5 Analysis of height response

The value of height response has been analysis and interpreted by using a half normal plot. Figure 7 shows the half normal plot graph for height response analysis. From the plotted graph the significant model effect which are point C, A and AC has been selected for the next evaluation using analysis of variance. However, point A model is automatically listed because to support the hierarchy. After selecting the model, the next step focused to the analysis of Variance ANOVA. Table 5 summarize the data of ANOVA analysis.

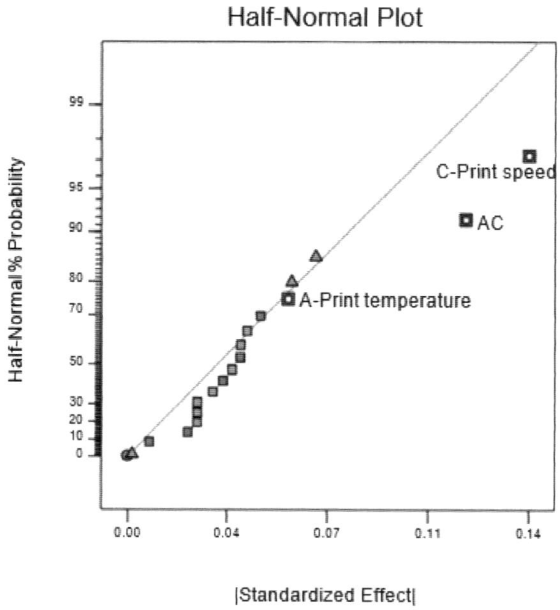

Figure 7: Half normal plot graph for width response

Table 5: ANOVA analysis of variance table for height response

Response : Height					
ANOVA for selected factorial model					
Analysis for variance table (Partial sum of square)					
Source	Sum of squares	df	Mean Square	F Value	Prob>F
Model	0.15	3	0.051	9.66	0.0009*
A-print temperature	0.013	1	0.013	2.54	0.1315
C-Print speed	0.081	1	0.081	15.44	0.0013
AC	0.058	1	0.058	10.99	0.0047
Curvature	0.004572	1	0.004572	0.87	0.3651
Residual	0.079	15	0.005242		
Lack of Fit	0.054	12	0.004498	0.55	0.8042**
Pure error	0.025	3	0.008221		
Cor Total	0.24	19			

3.6 Confirmation run

For confirmation run, suggested optimum run was repeated 5 time in order to verify the optimum parameter setting for PCL/PLA 3D printed cube by using fuse filament fabrication process. By using Flashprint software, the printed temperature are set to 136°C, deposition of height 0.12 mm, print speed 20mm/s and travel speed 22mm/s. In order to get the eligible test, the result from actual versus theoretical test must be in the margin of error less than 15%. From Figure 8 below, it can be seen that all the average error recorded for the length, width and height are below 15% which is acceptable.

Figure 8: Confirmation run result for (a) length, (b) width and (c) height

Run	Length of PCL/PLA 3D printed sample		
	Actual (mm)	Predicted (mm)	Error (%)
1	19.9985	20.000	0.0075
2	20.0525		0.2625
3	19.9470		0.265
4	19.9895		0.0525
5	20.0925		0.4635
Average error (%)			0.2102

(a)

Run	Length of PCL/PLA 3D printed sample		
	Actual (mm)	Predicted (mm)	Error (%)
1	20.0760	20.000	0.380
2	20.0090		0.045
3	20.0270		0.135
4	20.0470		0.235
5	20.0610		0.305
Average error (%)			0.220

(b)

Run	Length of PCL/PLA 3D printed sample		
	Actual (mm)	Predicted (mm)	Error (%)
1	10.1080	10.338	2.225
2	10.3665		0.275
3	10.1600		1.722
4	10.2665		0.691
5	10.4445		1.030
Average error (%)			1.1886

(c)

4.0 Conclusion

From this studies a sample size 2 x 2 x 2 cm in cube are successfully printed from PCL/PLA composites material using Fused Deposition Modelling (FDM) method with the aid of DOE. From the experiment, Analyzed of Variance (ANOVA) for length response, the significant effect that have been identified are print temperature (A), print speed (C) and interaction between deposition of height with travel speed (BD). Meanwhile for width response, the significant effect that have been identified are print temperature (A), print speed (C), interaction of print temperature with print speed (AC) and interaction of print temperature with travel speed (AD). For the height response, the significant effect are print temperature (A), print speed (C) and interaction of print temperature with print speed (AC). From combining analysis, the most significant effect that influence the dimensional accuracy of the PCL/PLA printed sample are print temperature and print speed. From the confirmation run the optimize parameter has been tested is 136°C print temperature, 0.12mm deposition of height, 20mm/s print speed and 22mm/s travel speed and by measuring the optimize parameter dimension, the value of error are below than 15% which is acceptable. This prove that PCL/PLA biocomposite can be printed using additive manufacturing technique. Further work have to be carried out in order to achieve the perfect printed part.

IOP Conf. Series: Materials Science and Engineering **607** (2019) 012001 doi:10.1088/1757-899X/607/1/012001

Acknowledgments

The authors wish to thank the Universiti Tun Hussein Onn Malaysia (UTHM) for the financial support to this work.

Reference

[1] Tempelman, E, Shercliff, H, and Eyben, B N V 2014 *Manufacturing and Design: understanding the principles of how things are made*. Butterworth-Heinemann.

[2] Majid Moayedfar, Ahmad Majdi Abdul Rani, S Emamiana, Ali Haghighizadeh and NahidJafari 2016 Manufacture of medical orthopaedic implant using computed Tomography imaging and rapid prototyping. *Journal of Engineering and Applied Sciences*. **11(1)**: 177-180.

[3] Chaim I A, Sabino M A, Mendt M, Müller A J and Ajami D 2012 Evaluation of the potential of novel PCL– PPDX biodegradable scaffolds as support materials for cartilage tissue engineering. *Journal of Tissue Engineering and Regenerative Medicine*. **6(4)**: 272-279.

[4] Joaquium de Ciurana, Lidia Serenoa and Elia Valles 2013 Selecting process parameter in RepRap additive manufacturing system for PLA scaffolds manufacture. *Procedia CIRP* 5 152-157.

[5] Haq R H A, Marwah O M F, Rahman M N A, Haw H F, Abdullah H, Ahmad S, Arifin A M T, Hassan M F and Yunos M Z 2018 Mechanical Properties of PCL/PLA/PEG composite blended with different molecular weight (MW) of PEG for Fused Deposition Modelling (FDM) filament wire. *International Journal of Integrated Engineering* **10** 187-192

[6] Hassan M F, Saman M Z M, Sharif S and Omar B 2013 Evaluation of sustainability performance of product design element concepts using analytic hierarchy process. *Applied Mechanics and Materials* **15** 799-808.

[7] Abdul Haq R H, Wahab M S, Wahit, M U B 2016 Fused Deposition Modelling of PCL/MMT/HA biocompatible polymer nano-composites. *Advanced Structured Material* **61** 23-22.

[8] Arifin A M T, Abdullah S, Zulkifli R, and Wahab D A 2014 Fatigue Failures of differences behaviour on CSM/ Woven Roving Composite Materials. *Applied Mechanics and Materials* **471** 335-340.

Application of Design for Six Sigma (DFSS) In Sustainable Product Design: A Case Study on an Electric Kettle

Mohd Fahrul Hassan[1], Tillak R Chandra Mohan[1], Al Emran Ismail[1], Ishkrizat Taib[1], Salwa Mahmood[2], John Mbogo Kafuku[3]

[1]Faculty of Mechanical and Manufacturing Engineering, Universiti Tun Hussein Onn Malaysia, 86400 Parit Raja, Batu Pahat, Johor, Malaysia.
[2]Department of Mechanical Engineering Technology, Faculty of Engineering Technology, Universiti Tun Hussein Onn Malaysia, Johor, Malaysia.
[3]Department of Mechanical and Industrial Engineering, College of Engineering and Technology, University of Dar es Salaam, P.O. Box 35131, Dar es Salaam, Tanzania.

*Corresponding author email: fahrul@uthm.edu.my

Abstract: Sustainability is one the most crucial aspects currently focused by competitive companies who are coerced to consider their companies environmental, social and economic performance as requirements for product design. The intricate philosophy of designing physical objects, products, built environment and services to comply with the principles of social, economic and ecological sustainability is called sustainable product design. The main of this study is to improve product sustainability using Design for Six Sigma (DFSS) approach. This research primarily focuses on reviewing current sustainable product development approaches, developing a new framework to incorporate Design for Six Sigma into Sustainable Product Development and to conduct a case study on a product using this new framework. The chosen product is an electric kettle. An electric kettle is usually subjected to more extreme usage and physical damage than normal kettles. Most original internal components are very hard to come by in the replacement parts market making it a difficult product to repair and extend its life of use. The design tools used in this new framework are survey methodology, Quality Function Deployment (QFD), morphological chart and weighted rating method, SolidWorks 3D modelling and Sustainability Analysis besides DFM and DFA analysis software. Through applying the new framework under the DFSS approach a new electric kettle concept has been generated and it has been proven to be more sustainable than its predecessors proving that the new proposed framework is suitable for use in sustainable product development.

Keywords: Product design, sustainable development, design for six sigma, DFMA.

1. Introduction

Product design in engineering terms is a series of concurrent steps, from understanding customer needs to commercialization, used to create a product design [1]. Product designers often embody various possible design concepts, evaluating each unique idea to turn them into innovative products and inventions through a very systematic approach. Art, science and technology comes together when the product designer creates an innovative product that people can use. Global sustainability, a topic widely discussed in recent times, is heavily influenced by the product design sector as almost all products consumed by each individual are outputs of the product development process [2]. In brief, sustainability or sustainable development is that of which meets the needs of the present community without compromising the ability of future, advanced generations to meet their own personal needs [3]. However, one of the greatest challenges faced by the modern society of today is the issue of sustainability, encompassing all social, environmental and economic topics whereby it is overwhelming both in scale and the complexity [2, 4]. This problem is further complicated by the fact that the unprecedented global population growth and the desire to improve the quality of life drives the consumption rate of products and services to keep on going higher. To make this vision a successful one, engineers and designers need to come up with radically creative solutions under the ever-present restraints of social, economic, and environmental sustainability considerations.

In general, to overcome this present issue, a generic problem-solving methodology called sustainable product design is used. Sustainable product design emphasizes on producing minimal amount of waste during the products entire cradle to grave life cycle from the onset of production, marketing, distribution, use and right up to the disposal of the product [5]. It is also important to note that design decisions have a very substantial impact on sustainability, in particular of those made in the early product design stages [6, 7]. Though there are many efforts worldwide being done to counter this worrying issue, analyzing the sustainability of each and every different product alternative is the toughest task since such comprehensive tools are either not yet available or not economically viable to producers. Sustainable product design clearly focuses on skilful, environmentally conscious design to reduce negative impacts of a product throughout its life. Sustainability is based on balancing three principal objectives which are environmental protection, economic growth and social equity [8]. Sustainable design on the other hand is about rethinking how to meet the need for growth while at the same time reducing negative environmental and social impacts [9]. It is also important to note that the most common approaches to achieving a sustainable product is by literally changing the design and the material composition of a product [10].

Therefore, it can be said that a design tool capable of evaluating a product with regards to sustainability considerations and selecting the most sustainability-oriented design alternative is still lacking thus further investigations are required to fill these gaps. Hence, this research will focus on creating a framework to promote sustainability in a manufacturing environment with the widely used problem solving tool, Design for Six Sigma (DFSS). Design for Manufacturing and Assembly (DFMA) approach is also integrated in the DFSS methodology to facilitate product improvement which can further support the economic sustainability of the product.

2. Methodology

In this paper, the main methodology used and referred to as a guideline is DFSS. This methodology is used to make a better improvement of an existing electric kettle towards sustainable product development.

2.1 Design for Six Sigma

Design for Six Sigma (DFSS) is a very innovative methodology for designing and redesigning products or services according to customer requirements and expectations. This DFSS methodology is a systematic and rigorous method within which tools, training and data are used to optimize the design process to achieve six sigma performances whereby it focuses on getting it right at the first time [11]. DFSS complements the six Sigma improvement methodology but takes it a step further by weeding out flaws of the product and process during the initial design stages itself. DFSS is simply not a rework of lessons learned in Six Sigma but a profoundly different approach [12]. The focus on early design phases in DFSS are primarily to account for customer needs and expectations besides proper production processes. DFSS incorporates its own roadmap as to what DMAIC is to Six Sigma. DFSS also uses another roadmap called DMADV. This roadmap is abbreviated into the following phases: Define, Measure, Analyze, Design and Verify [13, 14]. The DMADV roadmap focus is on redesigning an existing product or process so that the shortcomings of the product or process are corrected immediately.

2.2 Integrating DFSS Approach for Sustainable Product Design

Methods of sustainable product design can be the core principle in all the traditional design methodologies in which the desired outcome is a sustainable product. With major criteria's that needs to be considered in sustainable product design includes a very wide range of topics like environmental, economic and societal phases, it is important that new tools and approaches are found to assess the selected sustainability criterions. Numerous sustainable product design tools have been proposed and developed over the years. Current tools vary significantly in the data presentation methodology and design process implementation. Based on the literature review on existing sustainable product development, a new approach for investigating an existing electric kettle towards sustainable product development using DFSS is outlined in Figure 1. There are five different phases in the DFSS methodology and each of the phases are attached to a design tool. The DFM and DFA analysis was done in the verify part of the DFSS methodology. This is to ensure that the designed part is economically sustainable and can improve the products sustainability with regards to its manufacturing and assembly.

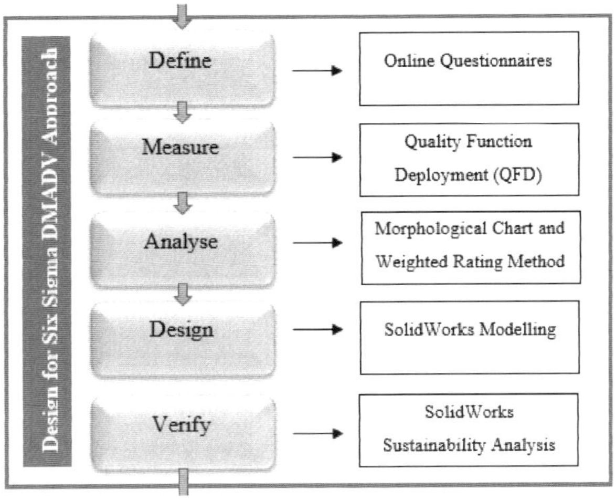

Figure 1. Proposed methods chosen for each DFSS Stage.

3. Case Study

In this section, a case study is conducted to illustrate the proposed approach. An existing electric kettle that is readily available in the market nowadays in the market was used as the case study and as a benchmark for the new proposed electric kettle design. The new conceptual design of the electric kettle must achieve key sustainability characteristics based on the three key elements of sustainability for example like customer satisfaction, life cycle orientation and significant design improvements. Design for Six Sigma (DFSS) is primary focused on designing products and service capable of meeting customer requirements or reaching Six Sigma quality methods.

3.1 Defining User Experience and Requirements for an Electric Kettle

In this DMADV approach, the first phase is known as the Define phase. The most suitable tool that can used in this to collect data is through a survey. Hence, an online survey form was created as a tool to help define the user experience and requirements on an electric kettle. It is of utmost importance that information and feedback from respondents are collected as their opinions on usage patterns can ultimately be the deciding factor in choosing or upgrading certain parts of the design. The respondents were 100 students who own an electric kettle where 45 of them were female and the rest were male students. Table 1 shows the sum of features fraction from feedback of respondents, there are total of the scores from respondents for each feature in this survey. Based on the data obtained from the survey, the respondents clearly want the conceptual design to be adjustable to use of temperature controls, energy efficient, safe enough to prevent hot water scalding and use non-toxic materials in the design of the new electric kettle.

Table 1. Sum of Features Fraction from Feedback of Respondents

100 Respondents	**Social (Users)**					**Economic**		**Environment**			
	Human Design: Ergonomics	Durability: Long lasting	Controls : Adjusts to Use	Safety : Scald Prevention	Maintenance: Easy to Clean	Maintenance : Common Components	Maintenance: Wide accessory and replacement hardware support	Energy: Efficiency	Material: Low Impact Materials	Material: Non Toxic	Material : Design for Recycling
Total	402	370	431	412	391	340	371	427	389	409	406
Average	4.02	3.70	4.31	4.12	3.91	3.40	3.71	4.27	3.89	4.09	4.06
Rating	0.092	0.085	0.100	0.095	0.090	0.079	0.085	0.098	0.089	0.094	0.093
Total Rating			0.462			0.164			0.374		

3.2 Measuring Respondents Satisfaction and Identifying Design Specifications for New Electric Kettle Design

In DFSS's DMADV approach, the second phase is called the Measure phase. The most desirable tool for this phase is Quality Function Deployment (QFD). The primary focus of QFD is to transform qualitative respondents' requirements on sustainable product development into quantitative parameters besides helping to deploy the function forming quality and methods of achieving the design quality into subsystems and component parts. QFD is designed to assist designer focus on characteristics of new or existing product from the viewpoints of market segment or technical development needs. QFD will transform respondents' needs or voice of customer into engineering characteristics for a product, prioritizing each product characteristic while setting the targets for product development at the same time. In the beginning a competitive benchmarking was done. Competitive benchmarking is the strategy of comparing the products of competitors. In this study, three existing electric kettle models from different brands in the market were chosen as competitive benchmarking. The three chosen kettle models were named as EK A, EK B and EK C. The analysis of the QFD was continued when the house of quality chart for the new electric kettle design was constructed. From this a target for the new product to achieve is obtained. Based on the analysis through the QFD, the highest total importance goes to product material which has a relative weight of 14.6%. Hence, the products material must be chosen properly according to customer needs and it must have a lesser impact on the environment. In second place is number of parts with a relative weight of 13.04%. This is related to the assembly and disassembly of the electric kettle. With lesser amount of parts, it is easier to assemble and disassemble. In third place is material life cycle with a relative weight of 11.75%. A much more sustainable material must be used in order to make the product at its end of life more recyclable.

3.3 Analyze and Selecting the Material to be Used in New Electric Kettle Design

The third phase of DMADV approach of Design for Six Sigma (DFSS) is the analyze part. In this specific phase of this research, a morphological chart and weighted rating method is used to generate new alternative concepts primary focusing on material choice and key functions. This analysis will aid tremendously with integrating key features into the conceptual design of the new electric kettle. With part material, reduction in number of parts and the material used life cycle key concerns of respondents in the survey hence it is of utmost importance of using materials which have a low impact on the environment throughout its entire use.

To ensure the new concept design is environmentally friendly throughout its entire lifecycle, selection of materials and other key components of the product is done by using a morphological chart and weighted rating method. Hence, through the application of these methods a greater emphasis to sustainability can be incorporated into the product design and its entire lifecycle. Table 2 shows the alternative combinations and list of parameters of new conceptual electric kettle design.

Table 2. Weighted Rating Method Scores of New Electric Kettle Design

NO	Criteria	Importance Weight	Concept Alternative 1		2		3		4		5	
			Rating	Weighted Rating	Rating	Weighted Rating	Rating	Weighted Rating	Rating	Weighted Rating	Rating	Weighted Rating
1	Human Design: Ergonomics	0.092	3	0.276	3	0.276	3	0.276	3	0.276	3	0.276
2	Durability : Long lasting	0.085	4	0.340	5	0.425	3	0.255	4	0.340	5	0.425
3	Controls : Adjusts to Use	0.100	4	0.400	4	0.400	4	0.400	4	0.400	5	0.500
4	Safety : Scald Prevention	0.095	3	0.285	5	0.475	5	0.475	3	0.285	5	0.475
5	Maintenance: Easy to Clean	0.090	4	0.360	5	0.450	3	0.270	4	0.360	5	0.450
6	Maintenance : Common Components	0.079	3	0.237	3	0.237	3	0.237	3	0.237	3	0.237
7	Maintenance : Wide accessory and replacement hardware support	0.085	3	0.255	3	0.255	3	0.255	3	0.255	3	0.255
8	Energy: Efficiency	0.098	4	0.392	5	0.490	3	0.294	4	0.392	5	0.490
9	Material: Low Impact Materials	0.089	4	0.365	5	0.445	3	0.267	3	0.267	5	0.445
10	Material: Non Toxic	0.094	4	0.376	5	0.470	3	0.282	3	0.282	5	0.470
11	Material : Design for Recycling	0.093	4	0.372	4	0.372	3	0.279	3	0.279	5	0.465
	Total	1		**3.658**		**4.295**		**3.29**		**3.373**		**4.488**

Based on the weighted sum ranking determined, the Combination 5 gained the highest total score of weights where it obtained a value of 4.488. The result of the weighted ranking method further supports the fact that the Combination 5 is the best design alternative from the five different combinations that was put up. Table 3 shows the final concept selection for the newly designed electric kettle.

3.4 Design and Modelling the New Electric Kettle's Conceptual Design using SolidWorks Software

The next phase in DFSS methodology is Design. Results from the survey which was thoroughly analyzed showed a desire from the respondents to have the product be more sustainable in terms of material used is non-toxic and has a complete life cycle whereby it can be recycled besides energy efficiency and reduction in the number of parts of the kettle. The new conceptual design has incorporated almost all of these desired characteristics with a special attention to detail. The 3D modelled detail drawings of the new electric kettle's conceptual design are modelled using SolidWorks 2015 software as shown in Figure 2. Table 4 shows the comparison of new electric kettle design and the existing electric kettle.

IOP Conf. Series: Materials Science and Engineering **607** (2019) 012002 doi:10.1088/1757-899X/607/1/012002

Table 3. Conceptual Selection of New Electric Kettle Design

NO	FUNCTION	SPECIFICATION
1	Kettle Body Material	Stainless Steel
2	Kettle Lid Material	Stainless Steel
3	Kettle Handle Material	Plastic
4	Kettle Base Material	Plastic
5	Heating Element Type	Flat 
6	Temperature Control	Electronic
7	Joining Methods	Screws
8	Screw Material	Stainless Steel
9	Screw Head Types	Flat 
10	Power Button Type	Toggle 
11	Power Button Shape	Round 

Figure 2. A 3D Drawing of New Electric Kettle Design in SolidWorks Interface

Table 4. Comparison of new electric kettle design and existing electric kettle

Existing Electric Kettle	Newly Designed Electric Kettle
Plastic Walls of the electric kettle body are not really heat resistant and can cause severe scalding issues.	New double walled design makes the hot water only come in contact with the inner layer making the outer layer less hot to touch.
Plastic material is not really environmentally friendly and has end of life issues of sustainability as because it cannot be fully recycled and ends up as plastic waste material. The plastic material also starts to leech chemicals into the water after some long usage. This effects indirectly the health of the person using the kettle.	The new stainless-steel design is focused at helping with the end of life recyclability issue of the electric kettle. Stainless steel is one of the most recycled material in the world and thus this reduced its environmental effect. This stainless steel also does not leech chemicals into the water thus making it a more socially concerned product.
The old electric kettle is set to a default temperature and sometimes over boils the water due to its thermostat sensitivity.	Water used need not be boiling hot to make a simple drink. The new electric kettle contains better electronic temperature control system which can boil water to a certain temperature as needed by the user.
The old electric kettle is not very ergonomically designed, and this can lead to people touching the wrong parts of the kettle and getting their selves a burn.	The new electric kettle design has an ergonomic handle so that the user comes in contact with the kettle at proper places. The lid is also designed flat with a small hole in between so that if ever the user touches the kettle that small hole will be where his hand will touch the lid.
The old electric kettle has a heating element that comes in contact with the water. This provides a means for the heating coils to get coated with a layer of scale that reduces the efficiency of the heating element as a whole which in turn translates to a higher electric power consumption. This scale slowly corrodes the heating element.	The new electric kettle has a flat heating element located at the bottom of the kettle. This new design does not make the heating element come in contact directly with the water making it more efficient for a longer time and reduces the chances of scale deposits on the heating elements corroding.

3.5 Verifying the Sustainability of Newly Designed Electric Kettle through a Sustainability Analysis in SolidWorks Software and DFMA Software

The main purpose of this analysis is to determine the environmental impact of a specifically designed product in terms of carbon footprint, air acidification, water eutrophication and total energy consumed throughout the lifetime of the product. The result of the environmental effects this product has is based on four major aspects of considerations which are material production, product manufacturing region, region for product use, end of life and transportation.

The sustainability analysis will be applied to evaluate sustainability of the new electric kettle design and compare its performance with the existing electric kettle design. By referring to the results of sustainability analysis component of the SolidWorks software, it can be clearly seen that there are differences on environmental impact level among the new and existing electric kettle design. Table 5 shows an in-depth comparison of the values obtained from the Sustainability Analysis Report. From Table 5, it can be seen based on the results each electric kettle is ranked by their environmental impact. Based on the result the new design has a lower impact on the environment when compared to the existing electric kettles design. Therefore, the new conceptual design of the electric kettle is able to bring benefits to the environment and can reduce the negative impacts on the environment. The main components of the electric kettle which is mainly using low impact materials like stainless steel helped this new design concept fare much better in this analysis.

Table 5. Sustainability Report of New and Existing Electric Kettle Design

	Sustainability Analysis	New Design	Existing Design
Environmental Impact	Carbon footprint (kg CO2e)	25	46
	Total Energy Consumed (MJ)	340	960
	Air Acidification (kg SO2e)	0.142	0.364
	Water Eutrophication (KgPo4e)	0.066	0.021
The least impact on Environment		1	2

Besides, Table 6 shows that, through the DFMA analysis it can be seen that the new electric kettle assembly is much cheaper to manufacture and assemble than the existing electric kettle. This reduction in time is mainly due to the fact that the setup of the machines in injection molding takes a longer time and the initial tooling investment in these injection molding is very high causing the parts to cost more although plastic is cheaper material. Plastic injection molding also has a high rate of reject items and other heat related forming issues like warping further adding to the waste of material thus reducing the sustainability of the product and increasing its production cost.

Table 6. Simplified Version of DFMA Report

	Existing Product	New Product
Total Cost per Product (RM)	56.14	37.56
Total Assembly Time (s)	253.46	63

4. Conclusion and Recommendation

In conclusion, it can be deduced that the main research objective of the study which is to improve product sustainability using Design for Six Sigma (DFSS) approach has been successfully achieved. The DFSS approach is used in this study to redesign an existing electric kettle that incorporates more sustainability related considerations in its build based on the customer requirements. Throughout each of the 5 phases of DFSS, from Define to Verify phase, the DFSS approach has provided means to incorporate rigorous design tools and processes to generate a better product design that has sustainability considerations in its core design process.

DFSS methodology serves as a very good approach to develop new products or redesign an existing product because this approach employs specialized tools as a guidance to ensure the outcome of design will achieve customer needs. For this study, a new electric kettle design was generated using the DFSS approach where this approach was proposed to reduce negative environmental impact and minimize products waste and also promote the use of low environmental impact materials. Replacing plastic parts with stainless steel makes more sense because it is recyclable and sustainable while being tough and have a better surface finish. The conceptual design of the new electric kettle has been verified to be much more sustainable that its predecessor, the existing electric kettle by using SolidWorks software.

In this study, the new conceptual design was only based on the materials used in the kettle and its impact on the environment. The other main elements in the electric kettle like the wiring and the efficiency of the heating coils was not really focused in this study. This field has a possibility to be further explored and if it is successful, it can further improve the sustainability of the electric kettle.

Acknowledgments

The authors wish to thank the Universiti Tun Hussein Onn Malaysia (UTHM) for the financial support to this work.

Reference

[1] Eggert R J 2010 Engineering design (2nd eds). High Peak Press, Meridian, Idaho.

[2] Hassan M F, Mahmood S, Saman M Z M, Sharif S and Sapuan S Z 2017 Application of Product Sustainability Evaluation Tool (ProSET) on Car Seat Design Configurations. International *Journal of Mechanical & Mechatronics Engineering* **17 (3)** 88-97.

[3] Holden E, Linnerud K and Banister D 2014 Sustainable development: Our Common Future revisited. Global Environmental Change **26** 130-139.

[4] Abdul Haq R H, Wahab M S and Wahid M U 2016 Fused deposition modeling of PCL/HA/MMT biocompatible polymer nano-composites. *Advanced Structured Materials* **61** 23-32.

[5] McLennan J F 2004 The philosophy of sustainable design journal/sketchbook. Kansas City, MO: Ecotone.

[6] Reid S E and Brentani U D 2004 The Fuzzy Front End of New Product Development for Discontinuous Innovations: A Theoretical Model. *Journal of Product Innovation Management* **21(3)** 170-184.

[7] Rahman M N A, Zakaria N H, Masood I, Adzila S and Nasir N F 2016 Risk assessment for assessing the subjective occupant seating discomfort related office works. Information (Japan) **19** 3025-3030.

[8] Hamid M K A 2014 Development of a Sustainable Integrated Process Design and Control for Chemical Processes. *Journal of Applied Sciences* **14(12)** 1277-1282.

[9] Vallero D and Brasier C 2008 Sustainable Design: The Science of Sustainability and Green Engineering. John Wiley and Sons, Inc.

[10] Hassan M F, Saman M Z M, Sharif S, Omar B, Zhang X, Badurdeen F and Jawahir I S 2013 Selection of Product Design Configuration for Improved sustainability using the Product Sustainability Index (ProdSI) Scoring Method. *Applied Mechanics and Materials* **315** 51-56.

[11] Shahin A 2008 Design for Six Sigma (DFSS): lessons learned from world-class companies. *International Journal of Six Sigma and Competitive Advantage* **4(1)** 48.

[12] Chowdhury S 2002 Design for Six Sigma: The Revolutionary Process for Achieving Extraordinary Profits Edition Dearborn Trade Publishing.

[13] Mast J D, Diepstraten G and Does R J M M 2011 Quality quandaries: Design for six sigma: Method and application. *Quality Engineering* **23** 204-211.

[14] Chowdhury S 2005 Design for Six Sigma: The Revolutionary Process for Achieving Extraordinary Profits. Kaplan publishing.

IOP Conf. Series: Materials Science and Engineering **607** (2019) 012003 doi:10.1088/1757-899X/607/1/012003

Effect of Filler Loading on Properties of Polyurethane/Clay Composites

Arifin, A. M. T.[1,*], Supri, A. G.[2], Hassan, M. F.[1], Rahman, M.N.A.[3], Haq, R.H.A.[3], Taib, I.[1], Adzila, S.[1]

[1]Department of Mechanical Engineering, Faculty of Mechanical and Manufacturing Engineering, Universiti Tun Hussein Onn Malaysia, Parit Raja, Johor, Batu Pahat, 86400, Malaysia.
[2]Faculty of Engineering Technology, Universiti Malaysia Perlis, 02600 Perlis, Malaysia.
[3]Department of Manufacturing, Faculty of Mechanical and Manufacturing Engineering, Universiti Tun Hussein Onn Malaysia, Parit Raja, Johor, Batu Pahat, 86400, Malaysia.

*Corresponding author email: mubarak@uthm.edu.my

Abstract. This paper discusses about the effected of filler loading on properties of polyurethane/clay composites. Polyurethanes materials have been developed strongly in the world, proving it to be one of the most versatile polymers. Segmented polyurethanes are very versatile materials, which can display properties ranging from very soft thermoset elastomer to strong. The objective of this study is to investigate the behaviour of polyurethane/clay composites, in term of strength, swelling behaviour and morphology properties. Polyurethanes based on diisocyanate molecule with diol were prepared by stoichiometric reaction 1:1 and 1:2 ratios of diphenylmethane diisocyanate (MDI) and polyethylene glaycol (PEG 400). In both ratios, 2%, 4%, 6%, 8%, 10% and 20% of clay were added as filler loading. Polyethylene glaycol (PEG 400) and clay are mixed together to become homogenous around 300 r.p.m. at room temperature for 10 minutes. After that, the solutions mixture with diphenylmethane diisocyanate (MDI) and are mixed again for 15 minutes around 300 rpm. The effect of adding clay was assessed by several testing method; compression test, swelling behaviors, spectroscopy infrared (FTIR) study and scanning electron microscope. Compression test result shows, with increasing percent of clay, the strength are increase for both of the ratio, 1:1 and 1:2. For swelling behaviors, from swelling index, result 1:1 ratio is better than 1:2 ratio. From infrared spectroscopy observation, the clay structure improved the hard segment of polyurethanes/clay composites. SEM micrograph of surface polyurethanes/clay composites observation shows, the structure of cell is homogenous; it is proved that chemical solution and clay are mixed together.

1. Introduction

Polyurethanes are one of the most versatile polymer materials, and it have been developed strongly in the world, today. Polyurethanes foams have a largest market among polymeric foams, estimated at nearly two billion kilograms in the US alone [1, 2]. Polyurethanes have an excellent abrasion resistance and the properties of both rubber and plastics. Polyurethanes are used in various application including as a thermal insulation, cushioning, buoyancy, energy absorption (packaging), etc. Due to low-density behaviour, these materials have been used as an interior panel for aircraft, structural shapes, shipping manufacture, sport equipment and composite tooling. However, polyurethanes also have disadvantages, such as low thermal stability and low mechanical strength [2].

IOP Conf. Series: Materials Science and Engineering **607** (2019) 012003 doi:10.1088/1757-899X/607/1/012003

Polyurethanes are made by the reaction of diisocyanate and a diol (R(OH)2), therefore the inclusion of various chain extenders allows a huge range properties to be achieved [3]. Polyurethane foams are available on thermoset or thermoplastic materials, from flexible (soft) too rigid or from tough to stiff elastomers [4]. Mechanical properties are important considerations for structural or semi-structural application as reported by Xia et. al. Therefore, due to polyurethanes, foams are formed by polymerisation method and foam blowing occur simultaneously, the polymer structure must build up rapidly to support the fragile foam, and not too fast to stop bubble growth [1]. Usually, polyurethane foams is used as a core for the sandwich structures in conjunction with two face sheet materials around it [4]. Core increases the energy absorption capability of the sandwich structure, but due to the weak the internal failure of sandwich structure is propagates through the core in the event of impact loading. Thus, it is important to improve the properties of thermal and mechanical properties of the core structures, which is polyurethane foams. To overcome these disadvantages, the development of nanostructured polyurethane (PU)/montmorillonite (MMT) composites in recent years [5, 6]. Effect of clay type, clay content and PU molecular structure on clay dispersion in thermoplastic PU nanocomposites have been studied, whereby it have been found that clays exchanged with long chain oniumions and good compatibility with polyol [1,5,6]. However, as reported by Tien et. al. a good dispersion of clay in the polyurethane matrix have been achieved through the modification of MMT with containing more than two hydroxyl group [7,8,9].

The literature regarding the use of nanoclay improved the thermal insulation and aging properties of polyurethanes foam [1]. The earliest motivation for the use clay in this research, in order to study the potential of low cost alternative materials to produce high performance composites structures, with an existing materials, which is diisocyanate and diol. In this study, polyurethane/clay composites, as well as foams were prepared by polymerisation method with different clay configuration. The effects of strength, swelling behaviour and morphology structures were investigated.

2. Methodology

2.1. Materials

Diphenylmethane-4, 4'-diisocyanate were used as a diisocyanate material and were obtained from Merck Schuchardt, Germany. The diol compound were obtained from Fisher Scientific, United Kingdom, Polyethylene glycol 400. Materials for clay were obtained from Zhejiang Fenghong Clay Chemicals Company LTD, with 13μm (Smectite Clay).

2.2. Polymerisation Methods

The diisocyanate and diol involved were weighed into the reaction in a 1:1 and 1:2, as shown in Table 1, with different composition of clay. The process of polymerisation, was started with combination of polyethylene glycol 400 and were added with clay (wt%) depends on the weight composition. After that, the mixer is used to mix both material to be homogeneous structure on the rotation speed of 300 rotation per minute (rpm). Diphenylmethane-4, 4'-diisocyanate was added after a few minutes of mixing process, with a little water was poured in the composition. The water was used as a blowing agent, whereby it is a substance that is incorporated into a mixture for producing foam (polyurethane). The process repeated until the required configuration of polyurethanes are formed. The polyurethane sample were cured at room temperature before the sample was cut out according to standard.

Table 1. Synthesis of polyurethanes with different composition of clay

MDI /PEG (v/v)	Clay (wt%)							Mixing (rpm)	Blowing agent (ml)
	Mix 1	Mix 2	Mix 3	Mix 4	Mix 5	Mix 6	Mix 7		
1:1 ratio	0	2	4	6	8	10	20	300	4
1:2 ratio	0	2	4	6	8	10	20	300	4

3. Results and Discussion

3.1. Mechanical Properties

Figure 1 shows the compression stress versus filler loading of polyurethane/clay composites for 1:1 ratio. It can be seen that, polyurethanes without clay have high compression stress compared with polyurethane/clay composites. The high compression stress, represent the hard segment is more influence on this configuration. The compression stress is decreased with increasing 2 wt%, 4 wt% and 6 wt% of clay. The compression stress at 2 wt% and 4 wt% is high compared with 6 wt% of clay in polyurethane/clay composites. However, if the 8 wt%, 10 wt% and 20 wt% of clay are added, the compressions stress is increased. It is proven that, with increasing more clay in configuration, the polyurethane/clay composites are return to become hard segment.

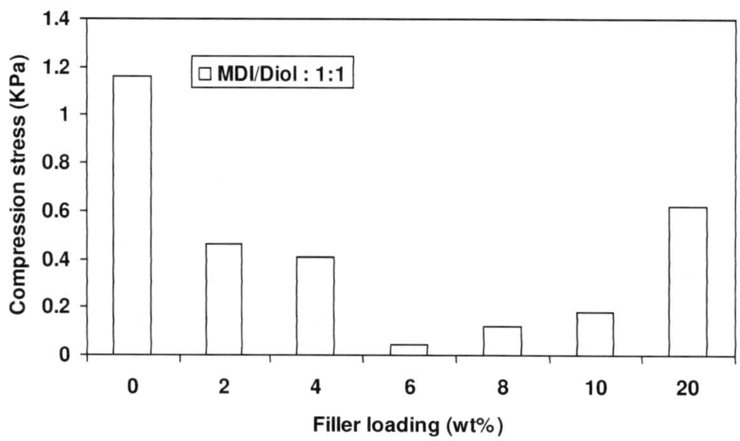

Figure 1. Compression stress versus filler loading of polyurethane/clay composites

Figure 2 shows the compression stress versus filler loading of polyurethane/clay composites for 1:2 ratio. The compression stress of polyurethane without clay are nil, due to soft segment are more influence on this ratio. However, with increasing clay as a filler loading, the compression stress increases. Based on the figure, it clearly shows with increasing of clay configuration for solution 1:2 ratio, from 2 wt%, 4 wt%, 6 wt%, 8 wt%, 10 wt% and 20 wt% of clay, the strength characteristic of polyurethane/clay composites are increases.

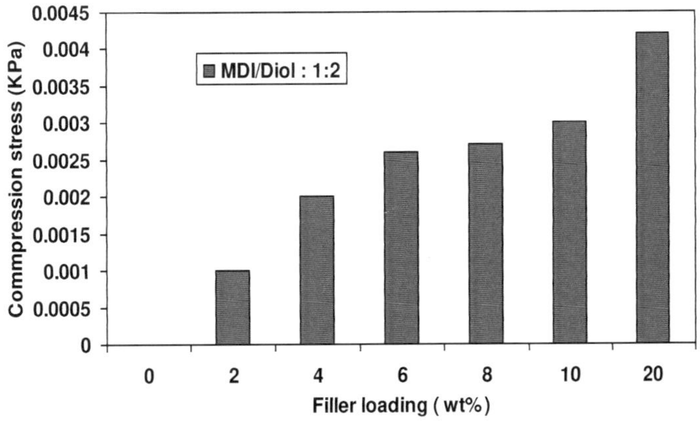

Figure 2. Compression stress versus filler loading of polyurethane/clay composites

3.2 Swelling Behavior Analysis

Table 2 (a) and (b) shows the data of swelling index polyurethane which immersed in chloroform at 24 hour for 1: 1 and 1:2 ratio. For 1:1 ratio, shown the polyurethanes without clay and 20 wt% have a lower swelling index compared with others polyurethanes/clay composites solution. However, the result of swelling index for 1:1 ratio is not constant as expected. For 1:2 ratio swelling index, it shows gradually increases with increasing clay on solution. This proves that with increasing percent of clay, the cross linking in polyurethane/clay composites decreases for 1:2 ratio, whereas the cross linking in polyurethane/clay composites increases for 1:1.

Table 2. Data of swelling index of polyurethanes, which were immersed in chloroform at 24 hour for ratio; (a) 1:1 ratio, (b) 1:2 ratio

Clay (wt %)	Swelling Index
0	11.23
2	12.03
4	15.70
6	19.14
8	18.93
10	18.61
20	14.08

(a)

Clay (wt %)	Swelling Index
0	7.08
2	11.31
4	13.94
6	15.34
8	16.48
10	17.04
20	21.74

(b)

3.3 Spectroscopic Infrared Analysis

Figure 3 (a) and (b) shows the spectra of polyurethanes with and without clay. As seen, the both figures, two peaks at 3339.5 cm^{-1} and 1528.5 cm^{-1} is N-H. The asymmetric stretching of the C=N bond is at 2366.3 $cm^{-1.}$ The asymmetric C-H stretch range to widen to 2920-2950 cm^{-1} and the C-O stretch is at 1228.3 cm^{-1}. However, in Figure 3(b) show the typical absorbing peak of Si-O bond exists at 1072.2 cm^{-1}. The absorption peak at 825.6 cm^{-1} is Si-CH_3. The characteristic bond of the polyurethane/clay composites at 1718.3 cm^{-1} as assigned to the carbonyl group (C=O). The peak appearing at 1734.5 cm^{-1} was attributed to the free carbonyl group (C=O). The correlation of both bonds (free carbonyl and carbonyl groups) is evidence of hydrogen bonding. The presence of hydrogen bonding in polyurethane improved the hard segment of polyurethane/clay composites.

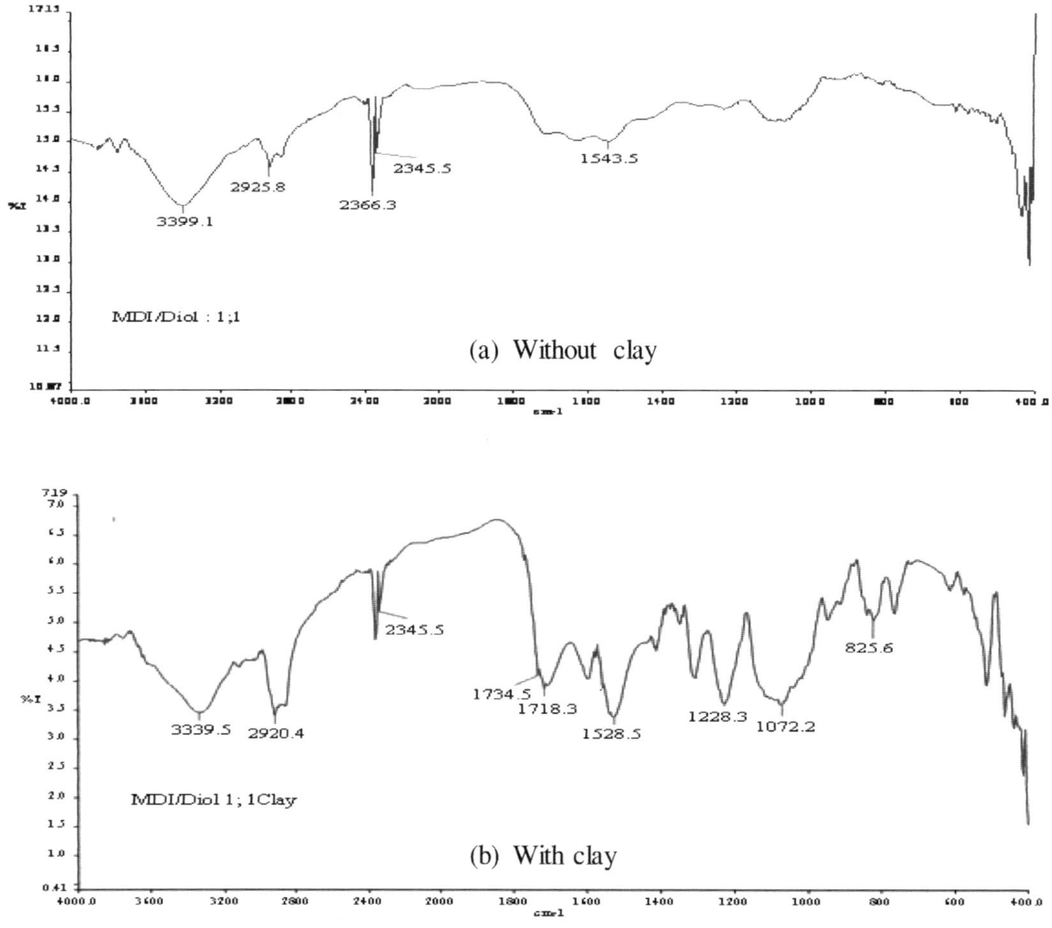

Figure 3. FTIR spectrum of polyurethane (MDI/ PEG: 1:1); (a) Without clay, (b) With clay

3.4 Morphology surface structure of Polyurethane

Figure 4 shows the SEM micrograph of the surface of polyurethane without and with clay for 1:1 ratio, are considered for comparison. Figure 4 (a) shows the SEM micrograph without clay. It can be seen that the structures is uniform with a small cell with porosity [9]. Even though, it have a porosity, but the strength is high compared to other solution due to stoichiometric ratio with hard segment characteristic. Figure 4 (b) shows the surface of polyurethane/clay composites, whereby the cell is agglomeration with a small and large cell.

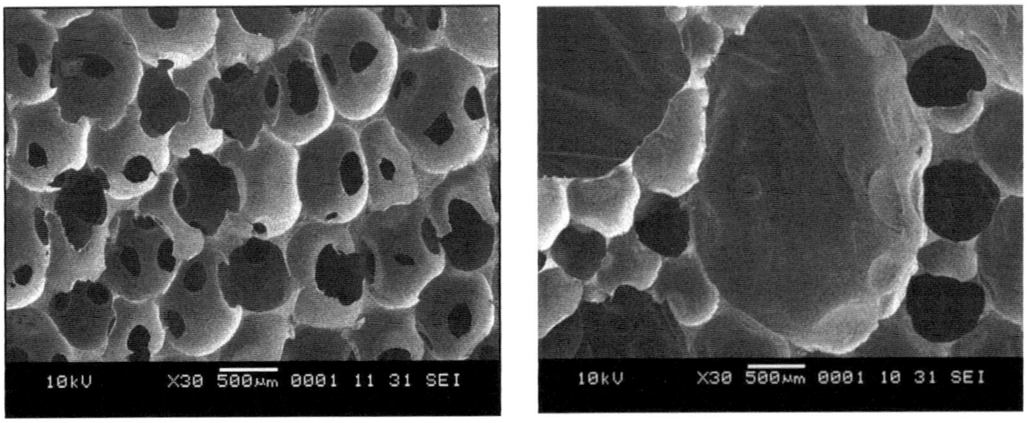

(a) (b)

Figure 4. SEM micrograph of surface of polyurethane; (a) Without clay, (b) With clay

4. Conclusion

The increasing strength is obtained from compression test. It was showed that with the increasing percent of clay, the strength increased for both of the ratio. However, the compression test for 1:1 ratio shown, the result is not constant, due to the stoichiometric reaction is not suitable with added clay as a filler loading in order to improve their strength. However, if more clay added in the solution, the strength also increased. In swelling index for swelling behaviors, the result of 1:1 ratio is better than 1:2 ratio. High cross linking in polyurethanes structures was observed proportional with increasing more clay as a filler loading. However for 1:2 ratio, it have more soft segment, the swelling index value is high due to absorption of more chloroform from the solution. Therefore, cross linking in polyurethane/clay composites is lower for 1:2 ratio. The infrared spectroscopy showed an additional of clay improved the hard segment of polyurethane/clay composites and correlates with improved in hydrogen bonding performance. From SEM micrograph surface of polyurethane/clay composites, it was showed the structures of polyurethane/clay composites cell are homogenous. In general, it is observed that chemical solution and clay can be mixed and shaped as polyurethane/clay composites.

Acknowledgement

The author would like to express their gratitude to Universiti Tun Hussein Onn Malaysia and Universiti Malaysia Perlis for supporting these research activities.

References

[1] Xia C, James L L, Tomy W and Christopher M 2005, Polyurethane/clay nanocomposites foams: processing,structure and properties, *Polymer* **46** 775-783.

[2] Krushna C P and Nayak P L 2012, Synthesis and characterization of polyurethane nanocomposite from castor oil-hexamethylene diisocyanate (HMDI). *Advances in Applied Science Research* **5** 3045-3052.

[3] Gareth J P, Emma J L and Christopher W G A 2010 The Effect of High Intensity Ultrasound on the Synthesis of Some Polyurethanes. *European Polymer Journal* **38** 1531-1536.

[4] Saha M C, Kabir M E and Jeelani S 2008 Enhancement in thermal and mechanical properties of polyurethane foam infused with nanoparticles. *Materials Science and Engineering* A **479** 213-222.

[5] Chang J H, and An Y U 2002 Nanocomposites of polyurethane with various organoclays: Thermomechanical properties, morphology and gas permeability. *Journal of Polymer Physics Part B* 670-677.
[6] Chen T K, Tien Y I and Wei K H 2000 Synthesis and characterization of novel segmented polyurethane/clay nanocomposites. *Polymer* **41** 1345-1353.
[7] Tien Y I and Wei K H 2001 High tensile property layered silicates/polyurethane nanocomposites by using reactive silicates as pseudo chain extenders. *Macromolecules* **26** 9045-9052.
[8] Arifin A M T, Abdullah S, Zulkifli R, and Wahab D A 2014 Fatigue Failures of Differences Behaviour on CSM/Woven Roving Composite Materials. *Applied Mechanics and Materials* **471** 335-340.
[9] Rafiquzzaman M, Abdullah S, and Arifin A M T 2015 Behavioural Observation of Laminated. *Journal of Fibers and Polymers* **16** 640-649.

IOP Conf. Series: Materials Science and Engineering **607** (2019) 012004 doi:10.1088/1757-899X/607/1/012004

Frame work for the Development of Occupational Safety and Health Assessment Model for Libyan Iron and Steel Industry

Alsaddeeq Basheer Yousuf Blaou[1]*, Mohd Nasrull Abdol Rahman, Mohd[1] Fahrul Hassan[1], Ibrahim Ahmed Badi[2], Tijjani Abdullahi[1]

[1]Faculty of Mechanical and Manufacturing Engineering, Universiti Tun Hussein Onn Malaysia, 86400 Parit Raja, Batu Pahat, Johor, Malaysia.

[2]Faculty of Engineering, Misrata University, Libya.

*Corresponding author email: gd170048@siswa.uthm.edu.my

Abstract. The aim of this study is to find an efficient method for assessment of occupational health and safety management (OHSM) practices which focused on the complexity of the iron and steel industry. Indeed, there are several method used in process management assessment but unfortunately, it does not hold the interaction that can exist between the health and safety risks management process and the complexity of iron and steel industry. To remedy this disadvantage, this study proposed reducing the number of indicators down to several or major KPIs. In order to achieve the selection of small set of KPIs from the larger set of PPIs, a SMART criteria and AHP in the domain of multi-criteria decision making (MCDM) analysis is proposed to be employed for developing a model for the assessment of OHSM practices in Libya iron and steel industry. The enterprises' safety performance index would benefit from the new assessment model.

1. Introduction

In history, not too long ago, the industry was brutally affected by ferocious occurrences in form of health crises and natural tragedies. Most of such incidents dealt a huge impact on production and economic activities of the industries. The repercussion of these mishaps are out of sequence and inconspicuous. For this reason, the occupational health and safety management in the industries has appeared to be an essential topic globally. Occupational risk has been defined as "a possible events whose unfavorable consequences are difficult to accept or are even unacceptable" [1]. In recent years, study on OHSM evaluation has become an essential area for research for the reason that occupational risks were continuously happening in the industries [2]. Iron and steel industries are among the occupations that are labelled with high-risk in the present era. Due to the blend of many reasons, which may include the high-risk working environment (which is typical of iron and steel industry) and low education level of unskilled menial workers, [3, 4].

The steelmaking process is very long and multifaceted. Such a complex enterprises normally go along with occupational health and safety risk, as well as the management of risk issues that emanates at the stage of iron making through the final stage of steel rolling. The risk management influences the entire system through marketing, information and inventory, procurements and transportation, which is because of the immense uncertainties created by the variance in the midst of decision-makers with different background and interest, policy strategies and the influence of marketing that might surface unavoidably in such an intricate organization.

It was rightfully established that the best and a well effective approach to manage risk is to improve safety performance. This according to Fung, et al. [5] should be geared towards preventing accidents as well as reducing uncertainties prior to its occurrence. Therefore, safety risk analysis can be viewed as the basis upon which concept and activities of safety management is built. Thus, "safety performance" assessment has turn out to be a serious task, which fashioned as fragment of safety management systems [6-8]. Based on the aforementioned, it is clear that in order to improve safety performance in the dynamic industrial set-up, safety personnel and specialists are the important figure to conduct thorough assessment in the field [9, 10]. Thus, it is imperative that safety personals' have a very good understanding and experiences on what safety assessment tools and models may affect the integrity and reliability of the performance assessment.

Previous studies concerns on some arrears like occupational risk factors or OHSM associated with construction and allied industries [4, 5]. Thus, studies related to "occupational health and safety" risk and risk management on iron and steel workplace are rare. Therefore, this study sort to build up on the need for effective OHSM in iron and steel industries through effective assessment of the OHSMs by developing an efficient and reliable assessment model suitable for iron and steel making industry. Therefore, criteria for effective model development have been put forward.

2. Literature review

In this section, the study attempt to create an impression about the nuts and bolts of safety and health management in accordance with the literature reviewed. The discussion will therefore include: basic definition of safety and health, safety and health management and the assessment method.

2.1. *Occupation health and safety*

Health and safety are part of "well-being", which in accordance with the dictionary definition, "well-being" is "welfare" [11]. It is therefore correct to say that, health and safety make-up a strong feature of the workers' welfare that have received quite a good attention for a period of time, especially as it relates to occupational management. "Health" in the words of Mathis and Jackson [12] is "a general state of physical, mental and emotional well-being". Thus, an individual that is "free of illness", psychological or emotional complications that mar his/her average human deeds is considered a healthy person. Therefore, the usual health management activities in industries and firms attempt to preserve the "overall well-being" of individual employees. Similarly, safety is the protection of the physical "well-being" of the populaces [12]. This means that, the sole purpose of active safety programs in firms and industries is for inhibiting incidents or accidents in the work places. Thus, the objective of safety is mainly for protection of workers, tools and the facilities.

The concept of health and safety is mostly discussed as safety and health hazard in many of the reviewed literature. Sikpa [13] defines safety hazards in accordance to the early work of Cascio, Wayne (1986) "as those aspects of the work environment that have the potential of immediate and sometimes violent harm to an employee" While the author describe Health hazards as "those aspects of work environment that slowly and cumulatively (and often irreversibly) lead to deterioration of an employee's health". Cole [14] enumerated the common causes of health and safety hazard to include "physical and biological hazards, toxic and carcinogenic dust and chemicals and stressful working conditions".

The concept of health and safety hazard had a continuous development since the coming of International Labor Organization (ILO) 1959 and beyond. Many health and safety policies have emerged, usually containing a clear aims and stated objectives of an organization concerning safety and health. It is often designed as an organized structure that is in a form of a system and procedure for the personnel that were given the responsibilities of handling issues concerning safety and health. Thus, referred as the health and safety management.

2.2. Occupation health and safety management (OHSM)

The major purpose of safety management is to preserve and promote personnel's health and safety. Hence, "health and safety management" is referred as an organized and prearranged activity that is governed by efficient management with a major objective for controlling health and safety risks [15]. In some case, the terms "safety program" and "safety system" are used instead despite that the latter two terms stress less the role of management [16]. Similarly, Booth and Lee [17] indicate that the main aim of safety management is to interfere with the causative process resulting in incidents and accidents. This involves active knowledge of the system detecting both visible and latent risks. It is a total system ensuring proper planning and executing of safety activities and arranging follow up system. Since the major activities of safety management circled around risk analysis, safety training arrangements, accident investigations, near miss investigations, safety promotion and human reliability" [17]. Safety performance improvements shall be recognized through a sustainable reduction in occupational and safety process deviations that lead to a loss, injury or damage. Thus, assessment of the management of health and safety practices needs a regular evaluation of progress achieved and to acquire feedback and suggestions for what's working and what's not.

2.3. Assessment of occupation health and safety management

Performance assessment usually give emphasis to measurement of performance on a certain subject matter, which is usually given a particular improvement strategies [18]. Whenever improving performance is mentioned, the tough on how best the implementation of performance management will be achieved become very important. This is true because it assists towards providing a systems for making a long and short term decisions with high benefits [19]. A familiar contribution by Bititci, et al. [20] added that "performance measure" is often acknowledged to be an essential instrument commonly employed in performance management, because it leads to the acquisition of important data that is relevant for taking decisions that can positively influence the performance of an organization. It can be declared that performance measurement has been taken as a topic of considerable interest for a long period, and it was mainly employed to measure performance as it relates to finance – in the form of income and returns or auditing. Accordingly, the economic measures of performance is popular and has continued to be the individual scale used for measuring the success of an organization. Nonetheless, Porter [21] stated "that performance scales based on fiscal indices cannot handle topical changes in industries, especially because of the appearance of novel technologies together with strong competition". This and many more indicates the need for providing appropriate methodology for measuring the success or otherwise of OHSM.

2.4. Approaches to health and safety performance evaluation

Studies were carried out to unravel the reason hindering the effective functioning of OHSM in many companies. The results of such studies, lead to some proposed factors that are considered to be the bottleneck to having a functioning systematic OHSM system. Researchers like Arocena and Núñez [10], Biggs, et al. [22] mentioned the lack of commitment as a factor, while the early work of Salminen [23], saw that lack of knowledge as the inhibiting factor. Although, the report by Nordlöf, et al. [24] has revealed that at present, adequate awareness on OHSM has been achieved. Nonetheless, the lack of formalized routines for effective auditing and improvement of the safety and health management system has been reported by quite a large number of literatures reviewed [7, 10, 25]. Karltun [26], Nordlöf, et al. [27] commented that such a situation where formalized practice for assessment and improvement is neglected, it will translate to failure of the OHSM system.

Lamenting on the lack of formalized routines for effective auditing and improvement of the safety and health management practices in companies, Teo and Ling [6] reported that "In Singapore, there was no regulation to govern the way in which the safety management systems (SMS) policies are drafted by construction firms" as at the time of their investigations. And that, "There is also no standard protocol on how safety auditing is to be conducted." Similarly, Nordlöf, et al. [24] reported that "No generally established instrument to measure OHSM practices was found" during their studies on Swedish national data. In the other hand, Beriha, et al. [28] found that, safety and health has "always been a major concern in the Indian industrial setting." However, the author further stated that only a small number of studies apply a reliable method for prediction of accident and assessment of the safety and health practices. Other studies that supported this claim include [29-32].

Further studies indicates that, the main approaches for assessment of safety and health management system are; "the retrospective or lagging indicators approach" and "the prospective or leading safety performance evaluation approach". The "Retrospective or lagging indicators approach" has been criticized by Carder and Ragan [33], Cooper and Phillips [34] with failures of not revealing the cause-effect relationships that would drive system improvement, and having less predictive value and the high possibility for under-reporting of incidence. In the order hand, the "Prospective or leading safety performance evaluation approach" has demonstrated the potency of keeping in touch with the modern organizational and safety management styles because it make available the information needing from "incident-based measurement" and it disclose how well an organization is carrying out those activities that are responsible for inhibiting injuries and ill-health. Thorough review of contemporary studies on safety and health performance evaluation indicates that most studies in this area of interest are based on the "Prospective or leading safety performance evaluation approach". Hence, this proposed study will be in accordance with this approach also.

3. Frame work for assessment of health and safety practice

The Libyan "iron and steel industry" was established in 1979 and situated around the coastal area of Misurata which is 210 Km away from Tripoli. The company has the capability of producing 1,324,000 tons of liquid steel annually. The operation of the "iron and steel industry" depends largely on direct reduction of iron and steel scrap [35]. It is very certain that in the long and complex process of steelmaking, initiatives are accompanied by risk from the iron making to the final steelmaking. It was reported by Abusa [36] that, "the industry also suffers from high accident rates, which results in absenteeism, loss of productivity, permanent disability, and even fatalities".

Usually, the best productive approach for improving safety performance have to be by inhibiting accident occurrence and decreasing uncertainty prior to its happening [37, 38]. Accordingly, safety and health risk study is a basis on which "safety management" is structured. Hence, qualified the safety assessment as a serious act of safety management systems [39, 40]. In view of that, a review of methods and techniques that have been applied for assessment of safety and health management system was conducted and only the performance assessment methods that are typical of and very useful in the area of health and safety management practices were selected and discussed. Consequently, in line with the criterion of a holistic approach and in accordance to the recommendations by [41-44]. The study ensure the development of PPI subsets for individual components of SH management system in accordance to the description in Figure 1. The preliminary result which is in agreement with Sgourou, et al. [44], shows a considerable large number of proactive performance indicators (PPIs) with quite huge internal structures were developed.

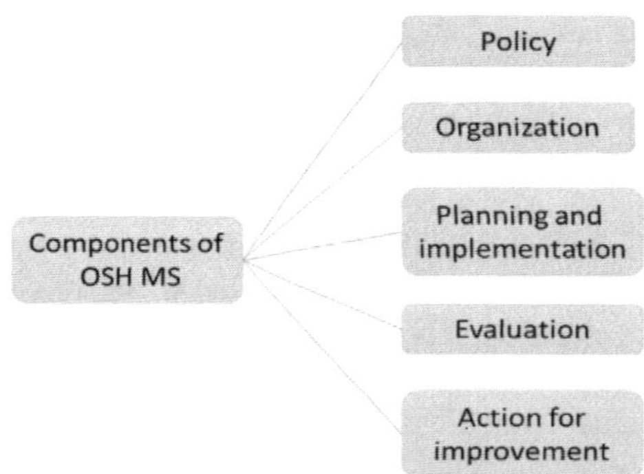

Figure 1. Components of SH management system for the development of PPI subsets.

The challenge as described by Podgórski [45] is the complexity of the measurement system due to bulkiness of data that is needed to be gathered and processed, where more time and personnel to perform measurements is required. The need for KPIs selection became inevitable as recommended by Keeble, et al. [46], Chan and Chan [47], Change [48], Øien, et al. [49] specify how necessary it is for decreasing the number of indicators down to major KPIs. The SMART criteria shown in Figure 2 was employed in determining the set of features that are expected of a good KPI.

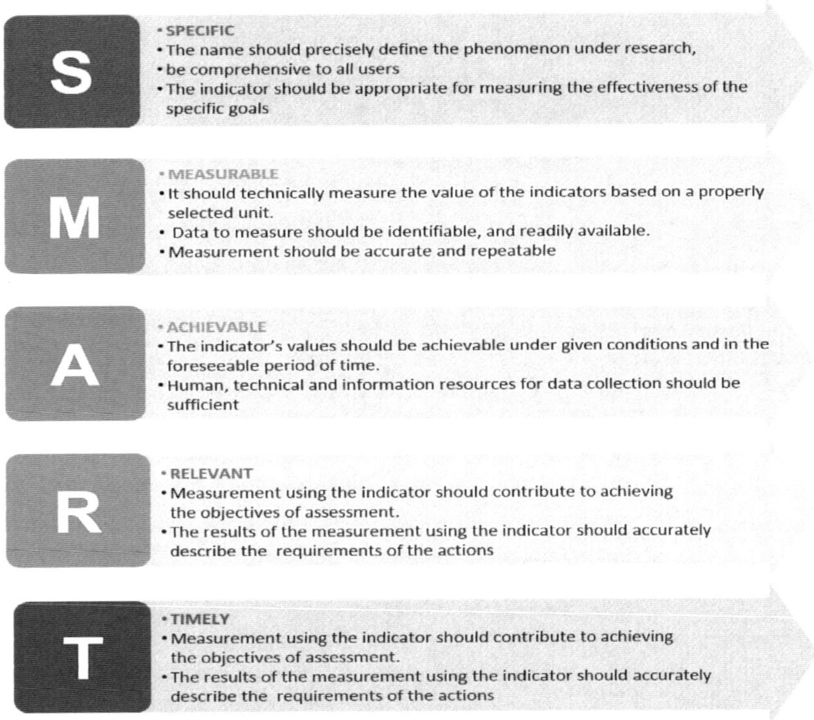

Figure 2. The SMART criteria for KPI selection.

In order to achieve a reduced number of KPI that can be suitably used in the evaluation of OSHM system, it is necessary to select the most important indicators from the large group of the already defined PPIs. An analysis of the available studies on the application of "Multi Criteria Decision Making" (MCDM) as a reliable methodology for selection of KPIs from the large group of PPIs indicates that many MCDM methodology can be employed for the selection. However, an analysis of the literature as presented in Figure 3, indicated that the "Analytic Hierarchy Process" AHP is among the commonly and the most used method in practice.

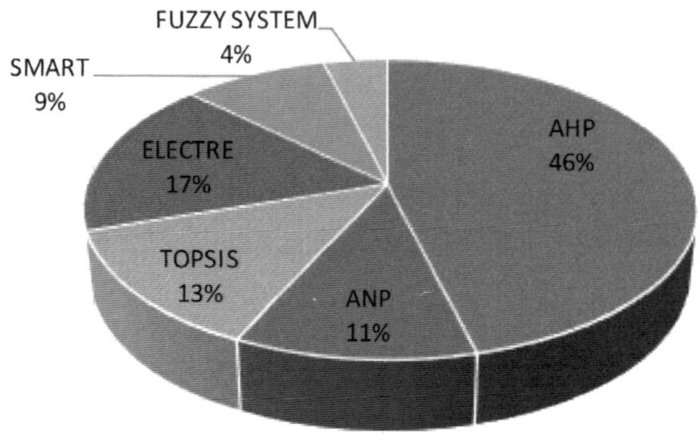

Figure 3. Applications of MCDM methods for KPI selection

The wider application of AHP in MCDM process was due to its moderate level of simplicity, coupled with the availability of relevant supporting software, and the prospect that it can be applied toward resolving decision issues concerning the economy, technology, leadership and areas of science. Therefore, this study choose to combine the decision making methodology of SMART and the hierarchical processing of AHP for developing the proposed model for the assessment of OHS practices in Libya iron and steel industry.

4. Conclusion

This research work appeared to be the first to study the active application of safety and health management in Libyan iron and steel industry, with particular consideration to the appraisal of the health and safety practices. The study's visible contribution is the presentation of various assessment techniques and the magnitude to which the techniques and philosophies are put to practice. As well as the contributions the techniques offered for managing health and safety risks effectively in the Libyan iron and steel industry. Important findings from the study is summarized below:

- Real status of safety and health management practices throughout the whole industry is uncertain. The absence of comprehensive blueprint towards realizing the full implementation of safety and health management practices is evident, with no standardized model or instrument for evaluation of safety and health management practices.
- The target to improve the Libyan iron and steel industry toward achieving adequate safety and health management practices is feasible only if the leadership, health specialists and the workers adjust to their responsibility and work together.
- The strategy for achieving effective safety and health management practices including sharing the vision of fulfilling the implementation process with all the related employees and ensuring the required resources for effective implementation of these practices.

Conclusively, significant amount of resources will need to be allocated for education and training on the significance of safety and health practices. Most especially, assessment of the safety and health management practices remained the only avenue to check on the effectiveness or otherwise of the system. Hence, having a robust assessment model that is unique for the Libyan iron and steel industry is very crucial and necessary.

References

[1] Tsai M C, Lai K H, Lloyd A E and Lin H J 2012 The dark side of logistics outsourcing–Unraveling the potential risks leading to failed relationships, *Transportation Research Part E: Logistics*

and Transportation Review, **48** 178-189.
[2] Olsson R 2007 In search of opportunity management: Is the risk management process enough?, *International journal of project management,* **25** 745-752.
[3] Liao C W and Perng Y H 2008 Data mining for occupational injuries in the Taiwan construction industry, *Safety science,* **46** 1091-1102.
[4] Niza C, Silva S and Lima M L 2008 Occupational accident experience: Association with workers' accident explanation and definition, *Safety science,* **46** 959-971.
[5] Fung I W, Tam V W, Lo T Y and Lu L L 2010 Developing a risk assessment model for construction safety, *International Journal of Project Management,* **28** 593-600.
[6] Teo E A L and Ling F Y Y 2006 Developing a model to measure the effectiveness of safety management systems of construction sites, *Building and Environment,* **41** 1584-1592.
[7] Hurst J and Jee-Hughes M 2001 Performance measurement and performance management in OECD health systems.
[8] Jung Y, Kang S, Kim Y S and Park C 2008 Assessment of safety management information systems for general contractors," *Safety Science,* **46** 661-674.
[9] Aksorn T and Hadikusumo B H 2008 Critical success factors influencing safety program performance in Thai construction projects," *Safety science,* **46** 709-727.
[10] Arocena P and Núñez I 2010 An empirical analysis of the effectiveness of occupational health and safety management systems in SMEs," *International Small Business Journal,* **28** 398-419.
[11] Dictionary C 2008 Cambridge advanced learner's dictionary," ed: PONS-Worterbucher, Klett Ernst Verlag GmbH.
[12] Mathis R L and Jackson J H 2004 Human Resource Management. International Student Edition. South-Western, a division of Thomson Learning, Thomson Learning is a trademark used herein under license," ed: Singapore.
[13] Sikpa F 2011 An Assessment of Occupational Health and Safety Practices on Job Performance at the Tetteh Quashire Memorial Hospital, Manopang Akuapan," ed: Kwame Nkrumah University Publishers, Ghana.
[14] Cole G A 2002 *Personnel and human resource management*: Cengage Learning EMEA.
[15] Burke M J, Sarpy S A, Smith-Crowe K, Chan-Serafin S, Salvador R O and Islam G 2006 Relative effectiveness of worker safety and health training methods," *American journal of public health,* **96** 315-324.
[16] Costella M F, Saurin T A and Macedo Guimarães L B de 2009 A method for assessing health and safety management systems from the resilience engineering perspective," *Safety Science,* **47** 1056-1067.
[17] Booth R and Lee T 1995 The role of human factors and safety culture in safety management," *Proceedings of the Institution of Mechanical Engineers, part B: Journal of Engineering manufacture,* **209** 393-400.
[18] Dwyer T 2013 *Life and death at work: industrial accidents as a case of socially produced error*: Springer Science & Business Media.
[19] Lebas M J 1995 Performance measurement and performance management," *International journal of production economics,* **41** 23-35.
[20] Bititci U S, Carrie A S and McDevitt L 1997 Integrated performance measurement systems: a development guide," *International journal of operations & production management,* **17** 522-534.
[21] Porter M 1992 The balanced scorecard: Measures that drive performance," *Harvard Business Review. USA.*
[22] Biggs S E, Banks T D, Davey J D, and Freeman J E 2013 Safety leaders' perceptions of safety culture in a large Australasian construction organisation," *Safety science,* **52** 3-12.
[23] Salminen S 1998 Why do small company owners think larger companies have fewer accidents? *Journal of Occupational Health and Safety Australia and New Zealand,* **14** 607-614.
[24] Nordlöf H, Wiitavaara B, Högberg H and Westerling R 2017 A cross-sectional study of factors

influencing occupational health and safety management practices in companies," *Safety science,* **95** 92-103.

[25] Holte K A and Kjestveit K 2012 Young workers in the construction industry and initial OSH-training when entering work life," *Work,* **41** 4137-4141.

[26] Karltun J 2004 Change processes and ergonomic improvements in small and medium enterprises, *Human Factors and Ergonomics in Manufacturing & Service Industries,* **14** 135-155.

[27] Nordlöf H, Wijk K and Westergren K E 2015 Perceptions of work environment priorities: Are there any differences by company size? An ecological study, *Work,* **52** 697-706.

[28] Beriha G, Patnaik B, Mahapatra S and Padhee S 2012 Assessment of safety performance in Indian industries using fuzzy approach, *Expert Systems with Applications,* **39** 3311-3323.

[29] Chen C L, Kaber D B and Dempsey P G 2000 A new approach to applying feedforward neural networks to the prediction of musculoskeletal disorder risk," *Applied ergonomics,* **31** 269-282.

[30] Ciarapica F and Giacchetta G 2009 Classification and prediction of occupational injury risk using soft computing techniques: an Italian study," *Safety science,* **47** 36-49.

[31] Fang D, Xie F, Huang X and Li H 2004 Factor analysis-based studies on construction workplace safety management in China," *International Journal of Project Management,* **22** 43-49.

[32] Sii H S, Ruxton T and Wang 2001 A fuzzy-logic-based approach to qualitative safety modelling for marine systems," *Reliability Engineering & System Safety,* **73** 19-34.

[33] Carder B and Ragan P W 2003 A survey-based system for safety measurement and improvement," *Journal of safety research,* **34** 157-165.

[34] Cooper M D and Phillips R A 2004 Exploratory analysis of the safety climate and safety behavior relationship," *Journal of safety research,* **35** 497-512.

[35] Iron L 1998 Steel Company, *Technical specification of products". Quality Control Department, Musrata,* Libya.

[36] Abusa F 2011 TQM implementation and its impact on organisational performance in developing countries: a case study on Libya.

[37] Hinze J and Gambatese J 2003 Factors that influence safety performance of specialty contractors," *Journal of construction engineering and management,* **129** 159-164.

[38] Abdullahi T and Abdullah S S 2015 Sustainability Considerations in Manufacturing and Operation Management," *International Journal of Scientific Engineering and Applied Science,* **1**.

[39] Tam C, Fung I W IV and Chan A P 2001 Study of attitude changes in people after the implementation of a new safety management system: the supervision plan," *Construction Management & Economics,* **19** 393-403.

[40] Zeng S, Tam V W and Tam C M 2008 Towards occupational health and safety systems in the construction industry of China," *Safety science,* **46** 1155-1168.

[41] A. Hale 2009 Why safety performance indicators?, *Safety Science,* **4** 479-480.

[42] Kjellén U 2009 The safety measurement problem revisited," *Safety Science,* **4** 486-489.

[43] Cambon J and Guarnieri F 2008 *Maîtriser les défaillances des organisations en santé et sécurité au travail: La méthode Tripod*: Lavoisier.

[44] Sgourou E, Katsakiori P, Goutsos S and Manatakis E 2010 Assessment of selected safety performance evaluation methods in regards to their conceptual, methodological and practical characteristics," *Safety science,* **48** 1019-1025.

[45] Podgórski D 2015 Measuring operational performance of OSH management system–A demonstration of AHP-based selection of leading key performance indicators," *Safety science,* **73** 146-166.

[46] Keeble J J, Topiol S and Berkeley S, "Using indicators to measure sustainability performance at a corporate and project level," *Journal of Business Ethics,* **44** 149-158.

[47] Chan A P and Chan A P 2004 Key performance indicators for measuring construction success," *Benchmarking: an international journal,* **11** 203-221.

[48] Change S 2003 Leading Performance Indicators: Guidance for Effective Use," *Step Change in*

Safety, Aberdeen.
[49] Øien K, Utne I B, and Herrera I A 2011 Building safety indicators: Part 1–theoretical foundation," *Safety science,* **49** 148-161.

IOP Conf. Series: Materials Science and Engineering **607** (2019) 012005 doi:10.1088/1757-899X/607/1/012005

Line balancing synchronization in powder coating workstation: A metal industry case study

F.H. Ho [*,1], M. Al-Haqeem Chee S. Abu1,[1] and Y. L. Woo[2]

[1] Department of Manufacturing and Industrial Engineering, Faculty of Mechanical and Manufacturing Engineering, Universiti Tun Hussein Onn Malaysia, 86400 Parit Raja, Batu Pahat, Johor, Malaysia
[2] Institute for Environment and Development (LESTARI), The National University of Malaysia, 43300 Bangi Selangor, Malaysia.

*Corresponding author email: fhho@uthm.edu.my

Abstract: In powder coating industry, product over-cured and poor surface finishing are both utmost issues faced by the manufacturer. One of the reasons is due to the poor synchronization between powder coating conveyor speed, number of operators in loading and unloading bay, and variety of products to be powder coated. This study is conducted in a metal industry company located at Johor with powder coating facility. Raw data were collected through documentation and observation such as complete operation activities of the powder coating process, the actual layout of the powder coating workstation. The raw data were then converted into simulation scenario using Arena software to determine the configuration between conveyor speed, number of operator, and consideration of three variants of product (small, medium, big) in order to achieve best synchronization. Next, the obtained findings were validated with the actual industry data. A balanced synchronized matrix is developed in the end of this study which could contribute to enable better understanding of line balancing in powder coating process. Perhaps the findings could be utilized to assist the practitioner in operation planning and productivity improvement

Keywords: Production efficiency, Line balancing, Powder coating, Lean manufacturing, Arena

1.0 Introduction

Achieving resources efficiency is one of the main goals for most of the manufacturers [1], for example practicing a good line balancing across the production. A well planned production operation will lead to higher productivity, whereby all of the involved workstations are well synchronized in term of machines involved and human speed [2]. Therefore, Line balancing is significantly important which commonly used to measure the workstation efficiency, particularly on the speed synchronization between related workstations through the work in progress (WIP) product [2, 3].

In powder coating line, metal products are coated in order to create protection against corrosion and mechanical wear or to give them the desired decorative finish [4]. Therefore, the powder coating technology is one of the most popular finishing methods available on the market today due to its good surface finish with protective coating [5]. However, not all of the powder coating line are fully automated in operation. In a complete powder coating workstation, it consists complex stations configurations with a combination of manual and automated activities. For example, human or labour are required in loading of semi-finished goods into the powder coating line, and unloading the coated product. Nevertheless, the common issues raised by the manufacturer such as line stoppage, product over-baked, and poor surface finish (color deterioration, uneven powder thickness). At worst case, powder coating layer peels off during the delivery to the customer. One of the reasons to explain this problem is due to the poor synchronization of the manual and automated operations in a powder coating line, especially the labour speed is not aligning with the conveyor speed. As a result, the product is over cured or left inside the curing oven for longer period due to line stoppage. Thus, a study is needed to simulate the scenario of the powder coating process and develop an evaluation matrix to suggest different configuration of man-machine-product.

2.0 Line Balancing

Imbalance of workstation are common happens when all involved workstation are varying in term of cycle time and speed. The common reasons are due to the complexity of the activity and the availability of resources. Complexity of activities in a workstation can refer to total steps or the number of activities involved to complete a task, whereas available resource refer to the number of manpower, equipment, tool, and machine to complete a task. In a common mean, in achieving a balanced line, all the participated workstations must be aligned or Synchronized in work-in-progress (WIP) productivity [6]. Example of a balanced production line, the task distributed across different workstations require to be in the same cycle time as shown in Figure 1 below.

Figure 1. A balanced production line with same cycle time across different workstation.

3.0 Powder Coating Process

Powder coating is a widely used painting process in many manufacturing industries. Powder coating is known as a dry finishing process that specify for a high quality, durable finish, maximized production, improve efficiencies, and simplified environment. According to Prasad et al. [7], powder coating is an advanced method of applying a protective and decorative finish to virtually any type of metal. The application of powder coating paint is widely used in automotive components, store fixtures, shelving, and electrical cabinet, metal furniture, and lawn or garden equipment.

Powder coating line is a complete system used to powder coat the products, where the products hanged up to the circular line manually or automatic. The product moved by conveyor all the way through several stages independently. The stages consist in line are loading area, pre-treatment, drying oven, powder booth, cure oven, and unloading area, and packing. The line has a variable speed that depends on the type of products so that the product can be hung and unload while the line is still running. Figure 2 below shows the lines with all stages in powder coating line [5].

Commonly in powder coating line, there are various problems occurring in powder coating line, for example, line stoppage due to in-alignment of human and machine speed, over-cured product issue that leads to poor quality of product. Both issue occurred due to difficulty to synchronize the activities that involve the type of product, conveyor speed and number of workforce. Therefore, in this study, aligning the human and machine speed in powder coating line is the main goal to enable no line stoppage.

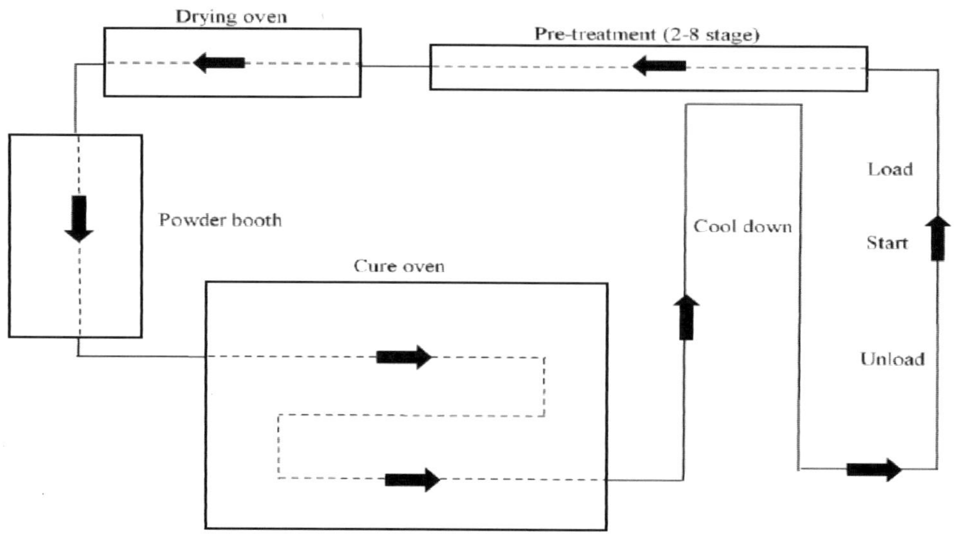

Figure 2. Complete powder coating line with its sub-stations (Du et al., 2016)

4.0 Methodology

This study was conducted based on case study methodology to obtain the industrial data [8,9]. Interview, observation and document analysis were conducted specifically in a company located at Simpang Renggam, Johor, Malaysia. The working procedures including cycle time were taken for the analysis based on different size and weight of the product (small, medium, big). As for the document analysis, the powder coating machine and workstation specifications were recorded. Next, the daily report which consists of the working hours, product variant to be powder coated, total manpower, and daily outputs were analyzed. These data were then transferred into simulation software according to the real working environment and the obtained speed such as cycle time of operator, machine conveyor speed, total man power, and the details operation of sub-station in the powder coating line. Arena simulation software (Ver. 12) was used to construct the modules of substations. The example of the powder coating line specifications is shown in Table 1. Figure 3 shows three different types of product to be powder coated.

Table 1. Powder coating line specifications.

Chain type conveyor powder coating line	Specification /Parameter
Speed of conveyor (max)	5m/min
Total conveyor length	262m
Number of spray gun	14 units
Number of spray booth	2 movable spray booth
Reusable powder material	Yes
Allowed dimension of product for powder coating	1400mm (high) x 400mm (width) X 12000mm (length)
Product weight per hanger (max)	120kg
Total substations	7

Figure 3. Three products type to be powder coated (a) Big size (b) Medium size (c) Small size

5.0 Result and Discussion

Powder coating process simulation is the output of this study. Therefore, a simulation scenario of powder coating line was developed (see Figure 4 below). The simulation parameters are followed the exact parameters of the case company such as the total length of the conveyor, number of hanger (to hang the product), distance between each hangers, and total substations.

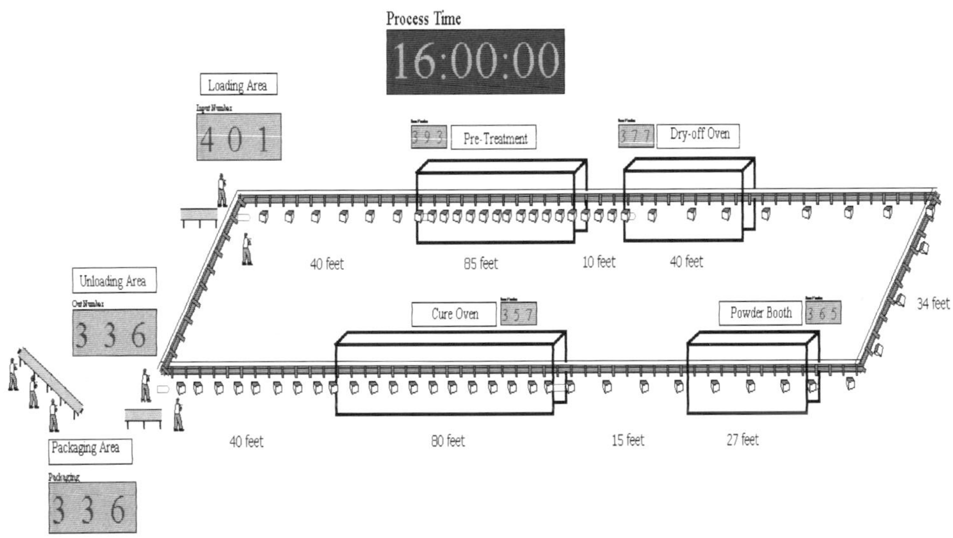

Figure 4. Powder coating process simulation using Arena software.

In this study, two types of evaluation matrix for powder coating line and validation were constructed (see Table 2 and Table 3 below). The matrix is constructed based on simulation of powder coating process. To obtain the simulation results, different configuration of trial runs was tested and synthesized in order to obtain the best synchronization between man-machine-products. The aims of the conveyor speed configurations are to avoid stoppage issue in powder coating line. Both of the evaluation matrix were constructed based on eight operation hours per shift were set at 8 hours which is same as the real working hours.

For dynamic conveyor speed (Table 1), three types of products were tested and analyzed. For small size product basic (e.g. base plate, bracing bar, supporting bar), the suggested best configuration of man-machine-product is 0.5 hanger distance to hang the product, while two labours is needed at loading and unloading area which could result in a total output of 434 pieces product. As for the medium size product such as shelving and slim beam, the suggested best configuration for conveyor speed 1.5 meter per minute is use of two labours at loading and unloading area. The estimated output per hour for this setting is 337 pieces of powder coated product. Next, for large sized product such as structured steel, standard beam. For conveyor speed at 1.0 meter per minute, the suggested number of manpower at product loading and unloading area is three persons due to heavier weight product. The estimated output per hour is 228 pieces.

For fixed conveyor speed scenario (refer to Table 2), the speed of conveyor is fixed at 1.3 meters per minute for three different types of product. For small sized product, two labours are needed at loading and unloading area with an estimated output of 1176 pieces. As for medium sized product, the optimum number of labour to be placed at loading and unloading area are three labours which could contribute to 326 pieces of outputs. Lastly, four labours are suggested at loading the unloading area for big sized product in order to avoid line stoppage due to unsynchronized speed between human and machine. A total 196 pieces output are simulated based on this configuration.

To summarize both of simulation condition, we can notice that different man-machine –product configuration could result to different production outputs in powder coating line. This scenario is unavoidable because the human need to be trained to align with the machine speed for achieving an optimum synchronization [10]. Thus, placement of the labour must be dynamic based on the product orders and the daily delivery requirement to optimise the resource usage and reduce potential poor product quality issue.

Table 2. Simulated best fit configurations of man-machine-product in powder coating line (with dynamic conveyor speed)

Product size (m)	Hanger distance (m)	No of labor	Labor speed (minute)			Conveyor speed (m/min)	Total output (pieces)
			Loading	Unloading	Packing		
Small (0.5Lx0.5Hx0.25W)	0.75	1	0.3	0.4	0.12	2.5	410
	0.5	2	0.2	0.3	0.09	2.5	**434 (124pcs/hour)**
	0.25	3	0.1	0.2	0.06	2.5	434
	NA	4	NA	NA	NA	2.5	NA
Medium (1.5Lx0.5Hx0.2W)	0.6	1	0.4	0.5	0.15	1.5	317
	0.45	2	0.3	0.4	0.12	1.5	**337 (48 pcs/hour)**
	0.3	3	0.2	0.3	0.09	1.5	337
	0.15	4	0.1	0.2	0.06	1.5	337
Large (2.0Lx0.5Hx0.2W)	1.0	1	1.0	1.1	0.33	1.0	142
	0.9	2	0.9	1.0	0.3	1.0	203
	0.8	3	0.8	0.9	0.27	1.0	**228 (33 pcs/hour)**
	0.7	4	0.7	0.8	0.24	1.0	228

Table 3. Simulated best fit configurations of man-machine-product in powder coating line (fixed conveyor speed: 1.3m/minute)

Product size (m)	Hanger distance (m)	No of labor	Labor speed (minute)		Total output (pieces)	Suggestion no. of operator
			Loading (fast,avg,slow)	Unloading (fast,avg,slow)		
Small (0.5Lx0.5Hx0.25W)	0.39	1	(0.2,0.3,0.4)	(0.25,0.35,0.45)	356	
		2	(0.2,0.3,0.4)	(0.25,0.35,0.45)	1176	**2**
		3	(0.2,0.3,0.4)	(0.25,0.35,0.45)	392	
		4	(0.2,0.3,0.4)	(0.25,0.35,0.45)	392	
Medium (1.5Lx0.5Hx0.2W)	0.39	1	(0.5,0.6,0.7)	(0.55,0.65,0.75)	NA	
		2	(0.5,0.6,0.7)	(0.55,0.65,0.75)	310	
		3	(0.5,0.6,0.7)	(0.55,0.65,0.75)	326	**3**
		4	(0.5,0.6,0.7)	(0.55,0.65,0.75)	326	
Large (2.0Lx0.5Hx0.2W)	0.45	1	(0.6,0.7,0.8)	(0.65,0.75,0.85)	131	
		2	(0.6,0.7,0.8)	(0.65,0.75,0.85)	157	
		3	(0.6,0.7,0.8)	(0.65,0.75,0.85)	170	
		4	(0.6,0.7,0.8)	(0.65,0.75,0.85)	196	**4**

****conveyor speed: 1.3meter/ minute***

6.0 Conclusion

Imbalance of production lines can cause low productivity and poor product quality. In this study, a complex powder coating process is chosen to be studied in term of the line balancing. Synchronization between man-machine-product were constructed. This study was conducted to simulate and reconfigure the man-machine-product speed in powder coating workstations. A guideline matrix was developed to demonstrate a different scenario of configurations in order to enable a powder coating line to operate steadily, particularly reduce the possibility of line stoppage and delay in the whole process of powder coating (loading, washing, rinsing, spraying, curing, unloading, packing). The matrix was fixed at a suitable conveyor speed, while dynamic parameters are the number of operators and variety of products. Thus, the suitable number of operators with estimated outputs are obtained (refer to Table 2 and 3). Validation of the matrix was conducted by comparing to the output data set collected from the selected company. Perhaps, the developed matrix can be applied to the industry with powder coating process, especially in operation planning to enable continuous movement of powder coating conveyor, which could avoid product poor quality issue such as product over-cured and uneven of coating thickness. For future research, more configurations of powder coating conveyor speeds to be studied in order to obtain a comprehensive matrix higher production efficiency.

Acknowledgments
The authors would like to thank University of Tun Hussein Onn Malaysia for funding this study, particularly under the Tier 1 research grant (Tier 1-H206).

Reference

[1] Ho, Fu H, Abdul-Rashid Salwa H, Raja Ghazilla Raja A, Woo Yoke L. 2019 Resources Sustainability through Material Efficiency Strategies: An Insight Study of Electrical and Electronic Companies." *Resources 8*, no. **2**: 117.

[2] Becker, Christian, and Armin Scholl 2006 A survey on problems and methods in generalized assembly line balancing." European journal of operational research 168, no. **3**: 694-715.

[3] Driscoll J and Thilakawardana D 2001 The definition of assembly line balancing difficulty and evaluation of balance solution quality." Robotics and Computer-Integrated Manufacturing 17, no. **1-2** 81-86.

[4] Mustonen Ulrika 2016 Pricing industrial powder-coating service: case: Mittametalli Oy.

[5] Du Zhongyan, Shaoguo Wen, Jihu Wang, Changle Yin, Dayang Yu, and Jian Luo 2016 The review of powder coatings. *Journal of Materials Science and Chemical Engineering 4*, no. **03**: 54.

[6] Zupan H and Herakovic N 2015 Production line balancing with discrete event simulation: A case study." IFAC-PapersOnLine48, no. **3**: 2305-2311.

[7] Prasad, Leena Kumari, James W. McGinity and Robert O. Williams III 2016 Electrostatic powder coating: Principles and pharmaceutical applications. *International journal of pharmaceutics 505*, no. **1-2**: 289-302.

[8] Yin Robert K 2006 Case study methods." Handbook of complementary methods in education research 3, 111-122.

[9] Noor Khairul Baharein Mohd 2008 Case study: A strategic research methodology. *American journal of applied sciences 5*, no. **11**: 1602-1604.

[10] Suchman, Lucy 2007 Human-machine reconfigurations: Plans and situated actions. Cambridge University Press.

IOP Conf. Series: Materials Science and Engineering **607** (2019) 012006 doi:10.1088/1757-899X/607/1/012006

Material removal rate and machining accuracy of electrical discharge machining (EDM) of Inconel 718 using copper electrode

S Ahmad[1,*], R N Chendang[2], A Supawi[3], M A Lajis[4] and M H K Zaman[5]

[1,2,3,4,5] Precision Machining Research Center (PREMACH), Faculty of Mechanical and Manufacturing Engineering, Universiti Tun Hussein Onn Malaysia (UTHM), 86400 Parit Raja, Batu Pahat, Johor, Malaysia.

*Corresponding author email: said@uthm.edu.my

Abstract. Electrical Discharge Machining (EDM) is a non-contact thermal machining process capable of machining conductive and semi-conductive materials. These capabilities bring to admission of machining materials that characteristically difficult to machine such as Inconel 718 due to its ability to maintain hardness at elevated temperature. Nickel based super alloy, Inconel 718 is one of the difficult to machine material that widely used in such high temperature application. The purpose of the study is to examine the Inconel 718 machinability by applied variable parameters such as peak current, pulse duration and repetitive used of electrode. To obtain the high performance measure in EDM, the selected peak currents were high where it is from 20 to 40A while for pulse duration is from 200 to 400µs. The performance measure that being evaluated are material removal rate (MRR), surface topography and machining accuracy. This research used Copper as electrode due to its high electrical conductivity and achieved highly in MRR. With the series of repetition used of electrode, the carbon layer known as black layer will formed at the surface contact of electrode. The relationship between the parameters and performance measure were more profound through the combination of machining parameters and black layer on the electrode. At high material removal rate, the surface topography of Inconel 718 look rougher with bigger sizes of nodules and craters produced. In the case of machining accuracy, it was found that, overall diameter overcut are less than 3.47%. The result shows that, even applying high peak current and pulse duration, the results of the machining accuracy is still acceptable.

1. Introduction

Nickel based super alloy, Inconel 718 is one of the material that machining by nonconventional machining such as Electrical Discharge Machining (EDM). Inconel 718 is a nickel-based high strength and high thermal resistance super alloy. Due to it has good resistance to fatigue, creep, corrosion at high temperature, Inconel 718 is widely used in gas turbine engines, steam turbines, nuclear power plants, rocket engines and spacecraft structural components [1]. This alloy is characteristically difficult to machine due to its ability to maintain hardness at elevated temperature which otherwise is very useful for hot working environment. Formation of complex shapes by this material along with reasonable speed and surface finish is not possible in traditional machining. Therefore, EDM is one of the most suitable processes to shape this alloy [2].

EDM is a non-contact thermal machining process capable of machining conductive and semi-conductive materials regardless of their hardness. With the used of EDM, highly accurate geometrically

IOP Conf. Series: Materials Science and Engineering **607** (2019) 012006 doi:10.1088/1757-899X/607/1/012006

complex and extremely difficult-to-machine materials can be machine compare with other conventional machining [3]. EDM is an extensively used to produce die and mould-making industry to generate complex cavities, aerospace, automotive industry and surgical components [4].

EDM is a thermoelectric process in which material is remove from work piece by erosion effect of series of electric discharges (spark) between tool and work piece immersed in dielectric liquid. The discrete electrical discharge will occurrence of sparks at the spark gap and as the results, the workpiece will be involved in material removal process. The eroded particles (debris) flushed away by dielectric liquid. Physical and metallurgical properties do not create any limitation for the materials to be machine on EDM, as there is no physical contact between tool and work piece [2].

One of the important decision in EDM when to cut the difficult-to-machine is the type of electrode. The electrode has to be high melting temperature, low wear rate and the cost is reasonable. Since the shaped electrode defines the area in which the spark erosion will occur, the accuracy of the part produced after EDM is high. After all, EDM is a reproductive shaping process in which the form of the electrode is mirror in the workpiece [4].

The machining accuracy becomes important when close tolerance components are required to be produce for space application and also in tools, dies and moulds for presswork, plastic moulding and die-casting. In the other hand, MRR for EDM operation is somehow is slower than with traditional machining methods, where the chips are produced mechanically [5]. Thus, this study has been perform in order to look for the effect of repetitive use of copper electrode by applying high peak current and pulse duration to the MRR and machining accuracy when EDM machining of Inconel 718.

2. Methodology

Before run the experiment, the top surface of workpiece, Nickel based super alloys, Inconel 718, flattened by grinding machine. The Copper electrode with diameter of 8mm flattened by lathe machine. The same electrode will be use repeatedly for the whole trials. The conducting of experiments using a standard Die-sinking EDM machine, Sodick AQ55L with custom-made working tank. This research aimed to investigate the roughing stage in EDM by applied 80% of duty cycle. The EDM experimental conditions and parameters summarized in table 1.

Table 1. Experimental conditions.

Parameters	Levels
Peak current, I_p (A)	20, 30, 40
Pulse duration, t_{on} (μs)	200, 300, 400
Pulse interval, t_{off} (μs)	Based on 80% duty cycle
Voltage, V	120
Electrode polarity	Positive
Dielectric fluid	Kerosene
Depth of cut	3mm

The weight of the workpiece and electrode for initial and after machining was measure using a digital weighing scale with four decimal point accuracy for MRR collected data. The microstructure of the surface workpiece after machining was taken using Scanning Electron Microscope (SEM), while for diameter measured of the machined hole and electrode was taken using flexible measuring machine with four decimal point accuracy. Figure 1 shows the experimental setup before running the EDM operation. The following equation was use in this research:

$$\text{Duty cycle} = \frac{\text{Pulse On time}}{\text{Pulse On time} + \text{Pulse Off time}} \quad (1)$$

$$\text{MRR(g/min)} = \frac{\text{weight}_{\text{before}} - \text{weight}_{\text{after}}}{\text{machining time}} \quad (2)$$

$$\text{Diameter overcut}\,(\%) = \frac{\text{diameter machining} - \text{diameter electrode}}{\text{diameter electrode}}\, x\, 100\% \qquad (3)$$

Figure 1. Experimental setup.

3. Results and discussion

Table 2 shows the experimental results for EDM of Inconel by using Copper electrode.

Table 2. MRR and machining accuracy result.

Peak current, IP (A)	Pulse duration, T_{on} (μs)	MRR (g/min)	Diameter overcut (%)
20	200	0.0591	2.77
20	200	0.0685	3.07
20	200	0.0720	2.62
20	300	0.0321	3.24
20	300	0.0317	2.74
20	300	0.0354	3.35
20	400	0.0216	3.47
20	400	0.0222	3.26
20	400	0.0216	3.31
30	200	0.1325	2.77
30	200	0.1433	2.61
30	200	0.1349	2.90
30	300	0.0973	2.83
30	300	0.1165	2.61
30	300	0.1232	2.78
30	400	0.0709	3.46
30	400	0.0818	2.59
30	400	0.0973	2.80
40	200	0.1685	2.99
40	200	0.1838	2.97
40	200	0.1784	3.03
40	300	0.1324	2.89
40	300	0.1590	2.97
40	300	0.1607	2.68
40	400	0.1213	2.92
40	400	0.1277	2.98

40	400	0.1241	2.92

3.1. MRR Analysis

The highest MRR value is 0.1838g/min on the second hole (40A, 200µs), while the lowest is 0.0216g/min on the first hole (20A, 400µs) and third hole (20A, 400µs) shown in Figure 2. This shows that the highest MRR own by the highest peak current in this experiment. Figure 2 shows MRR value increase proportionally with increasing the peak current with all pulse duration conditions. These phenomena happen to attribute the high peak current in large impulsive force towards workpiece during high frequency series of discrete electrical discharges. Further facilitates the melting and vaporization process, resulting faster machining time [2].

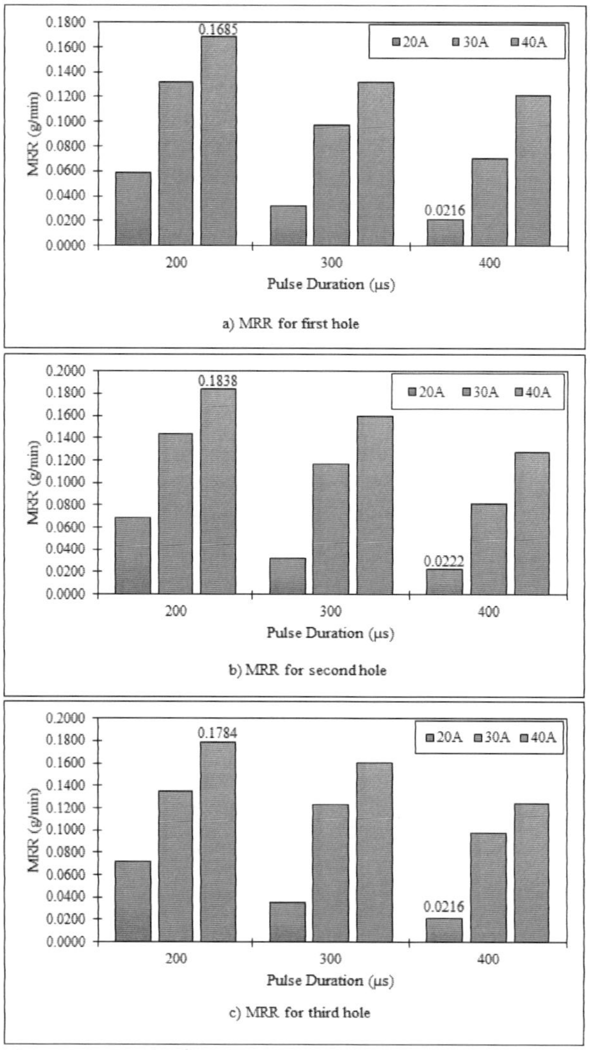

Figure 2. Effect of peak current and pulse duration on MRR by repetition used of electrode [(a) first hole, (b) second hole and (c) third hole].

As shown in figure 2, the highest MRR obtained by 200µs pulse duration for each hole as 0.1685, 0.1838 and 0.1784g/min respectively. As demonstrated, the MRR keep decrease as pulse duration increased from 200 to 400µs. this results proved that by continuously increasing the pulse duration does not necessarily improve the MRR. To illustrate this result, the higher pulse duration makes the longer extension of the plasma channel to collapse and resulting the bigger size of plasma channel. Further, its will ensue reduction on its energy density and heat energy per unit plasma [6]. Thus, by increasing pulse duration resulting decrease in MRR. It is notice that, the highest MRR obtained at second hole with 0.1838g/min compare to the first hole with 0.1685g/min with 9.08% of increment. Then, the MRR is slightly decrease at the third hole with 0.1784g/min. This is due to the formation of carbon layer known as black layer on the contact surface of electrode at second hole and third hole. When the thermal energy process through high frequency series of discrete electrical discharges between the electrode and workpiece happens, the molten material been evacuating on the surface of workpiece. Crater is form on the workpiece surface while on electrode surface, the forming of black layer occurred as react with dielectric fluid through pyrolysis occurring during the discharge plasma [7]. The results was corresponding with Rajesha et. al. [8], they found that with the high peak current, MRR could improve with assistant of carbon deposited on the electrode.

This could explain that black layer on the electrode are harder than electrode materials that resulting difficult erosion to the electrode base and improve in increasing MRR. However, the effect of carbon is complex and still not fully discover on the aspect of the plasma breakdown such ionization [9].

3.2. Surface topography analysis

In order to simplify the surface topography result, two different magnifications for each parameter were observe, respectively. Figure 3 shows three-dimensional SEM images for surface topography of Inconel 718 machine by EDM that indicate the highest and lowest MRR for first hole. It was notice that the surface topography characteristic are closely relate to the MRR and directly depends on the applied peak current. Referring to figure 3(a), at low MRR, the micro geometry characteristics of the surface shows the existence of craters, micro-cracks, nodules, voids, pork marks and globules.

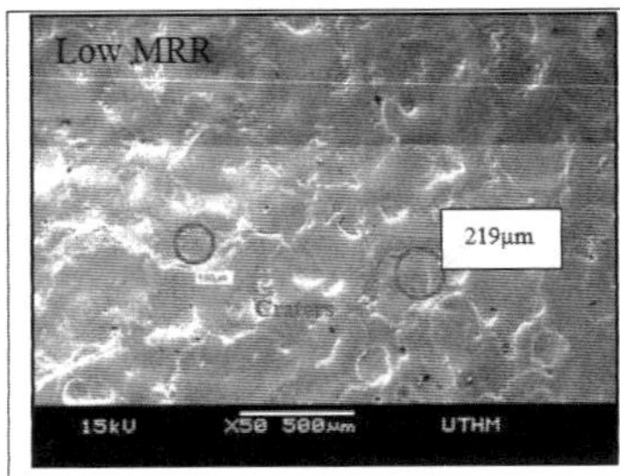

a-1) MRR=0.0216g/min; Ip=20A; t_{on}=400µs

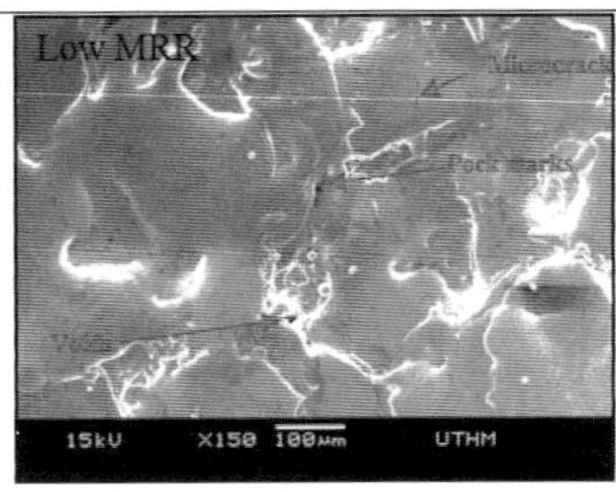

a-2) MRR=0.0216g/min; Ip=20A; t_{on}=400µs

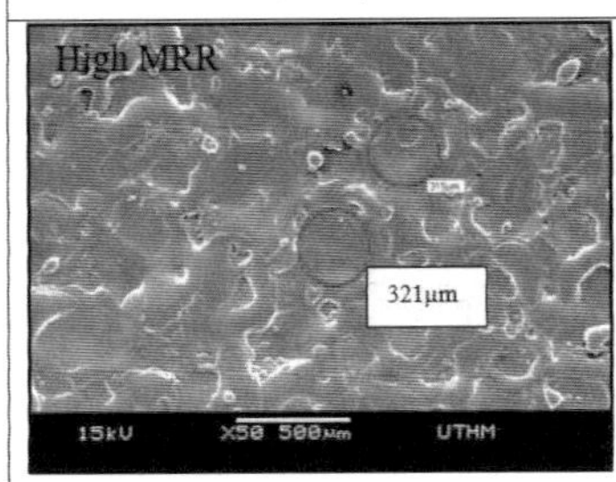

b-1) MRR=0.1685g/min; Ip=40A; t_{on}=200µs

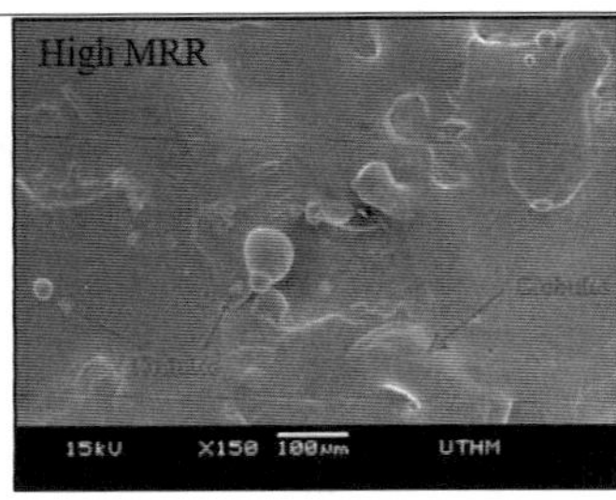

b-2) MRR=0.1685g/min; Ip=40A; t_{on}=200µs

Figure 3. Surface topography of Inconel 718 at first hole with different magnification.

However, in figure 3(b) at high MRR, the craters and nodules size were noticeable bigger than low MRR, which is for low MRR was 219µm, and the high MRR was 321µm. Besides that, the globules become thicker and the craters got deeper. This could be explain by low MRR the pulse duration was in

IOP Conf. Series: Materials Science and Engineering **607** (2019) 012006 doi:10.1088/1757-899X/607/1/012006

high setting resulting more material melted at long time and cause the molten material not swept away from the surface workpiece. Hence, the nodules and globules of the debris formed [10]. Besides, with the high peak current, the massive material melting made the nodules and globules become multi-layer. On the other hand, pork mark produced by the re-solidification of the materials that made the structure become sharp-edge-geometry [11].

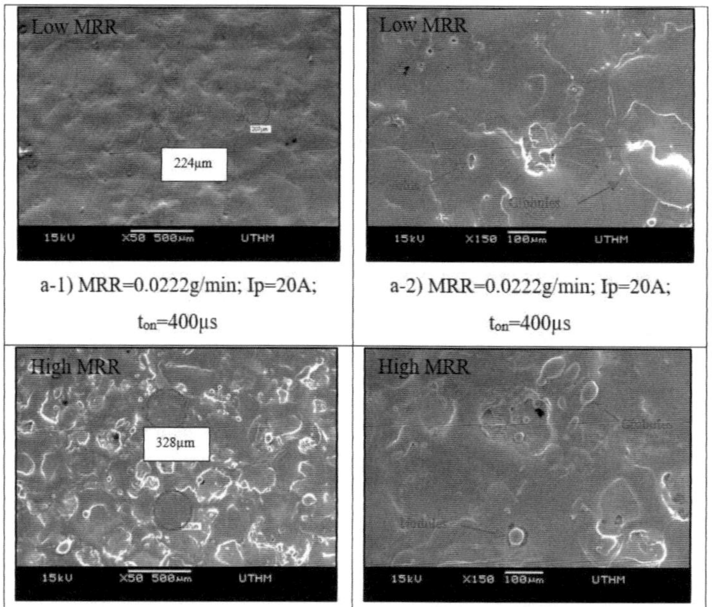

a-1) MRR=0.0222g/min; Ip=20A; t_{on}=400µs

a-2) MRR=0.0222g/min; Ip=20A; t_{on}=400µs

b-1) MRR=0.1838g/min; Ip=40A; t_{on}=200µs

b-2) MRR=0.1838g/min; Ip=40A; t_{on}=200µs

Figure 4. Surface topography of Inconel 718 at second hole with different magnification.

Based on figure 4, when electrode was use for second time with the same peak current and pulse duration, the surface topography at low MRR and high MRR produced slightly rougher surface with the bigger nodules and craters size than shown in Figure 3. The crater size for low MRR is 224µm and for the high MRR is 328µm, respectively. This can be explain that, by having black layer coating the outer surface of electrode produced higher thermal energy. This is due to the hardened shape of the carbon that increased the MRR and further increase the surface roughness. When peak current and pulse duration high, the deterioration of the surface texture on Inconel 718 composed of bigger size of craters and nodules that cause from the overlapping of melting materials during the successive electrical discharge in machining process. Furthermore, the micro-crack is cause from the high apply of temperature. This is due to the intern stress grow during the solidification progresses of the materials [12].

As shown in figure 5, both craters size for low MRR and high MRR at third hole were less compare to the second hole. The low MRR produce the size of craters 210µm and the high MRR produce 322µm. The craters on the surface were shallow than second hole. This is due to the low MRR value that directly affected the outcomes of surface topography. When low eroding process, the surface roughness is low due to the shallow craters that produces from the low impact of the discharge energy [12].

Furthermore, the craters size depends on the explosion of the spark occur during machining time. The craters become bigger and deeper when the peak current is high due to the spark intensity and high energy bombarded on the workpiece surface. However, the micro-cracks are not noticeable within low MRR and high MRR.

IOP Conf. Series: Materials Science and Engineering **607** (2019) 012006 doi:10.1088/1757-899X/607/1/012006

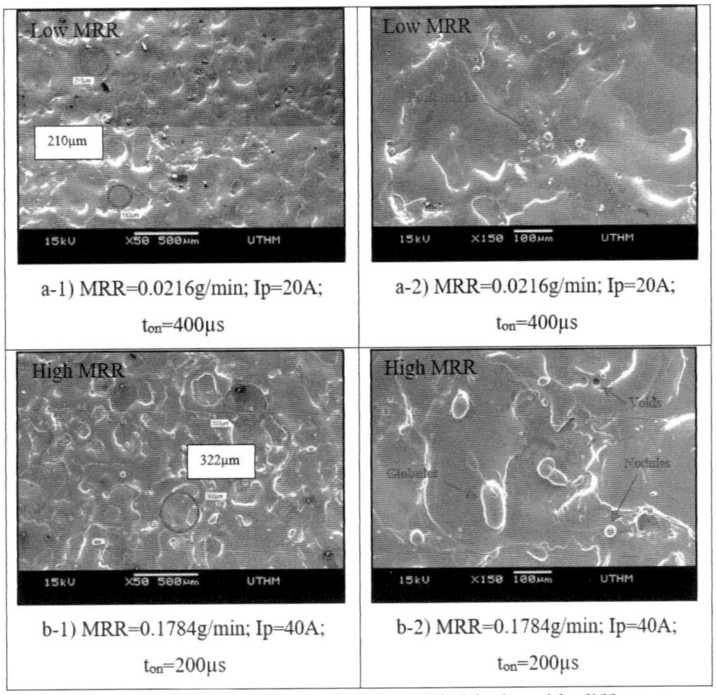

a-1) MRR=0.0216g/min; Ip=20A; t_{on}=400µs

a-2) MRR=0.0216g/min; Ip=20A; t_{on}=400µs

b-1) MRR=0.1784g/min; Ip=40A; t_{on}=200µs

b-2) MRR=0.1784g/min; Ip=40A; t_{on}=200µs

Figure 5. Surface topography of Inconel 718 at third hole with different magnification.

3.3. *Machining accuracy analysis*

In general, the less diameter overcut indicate the better machining accuracy. As mention earlier, machining may take place on the side space leading to diameter overcut. The highest diameter overcut is 3.47% on the first hole (20A, 400µs), while the lowest is 2.59% on the second hole (30A, 400µs) shown in figure 6. This shows that the highest machining accuracy obtained by lowest peak current and directly to the lowest MRR. Besides, the lowest machining accuracy owns by the average peak current and MRR. However, for overall trends of machining accuracy were fluctuated towards peak current and MRR.

The results shows that the machining accuracy were not attach to any parameters input. However, as seen in the graph, the diameter overcut slightly fluctuated by increasing peak current but still in consistent trend. Owing to the earlier discussion that by increasing peak current obtained high MRR, this can be explain that as MRR is high the high frequently spark produced were at the limit and leads to constant explosion to form crater. The crater produced were constant and leads to constant overcut of machining. The experimental results here may further show that as more heat transferred to the workpiece as the peak current increases, the more constant of diameter overcut. However, this experiment result contrast with Ghewade and Nipanikar [13] finding which the machining accuracy proportionally increase with peak current increase as at low peak current resulting low spark energy which indicate the smaller crater formed on the surface of workpiece, further resulting in good machining accuracy. It can be seen that the parameter used were smaller than this experiment conducted. This could explain that by using the smaller parameters indicate the proportional increase in machining accuracy. However, the result of MRR for this parameter used were not satisfied for roughing stage where objective of this research conducted. Therefore, from the machining accuracy result, it shows that by increasing the peak current indicate the constant machining accuracy.

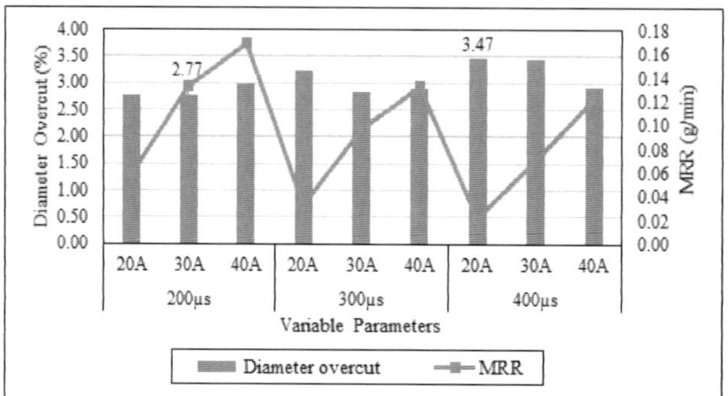

a) Diameter overcut against MRR for first hole

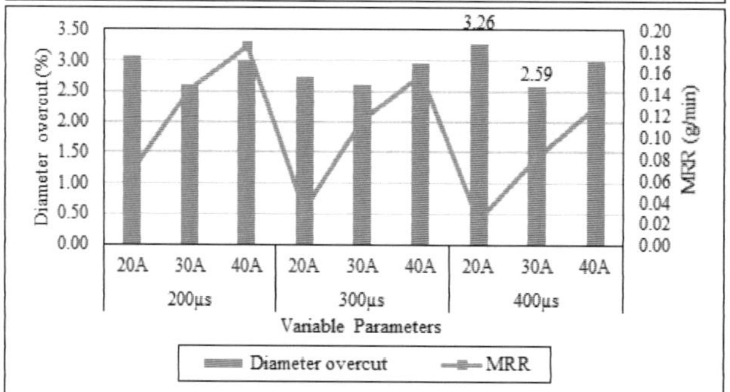

b) Diameter overcut against MRR for second hole

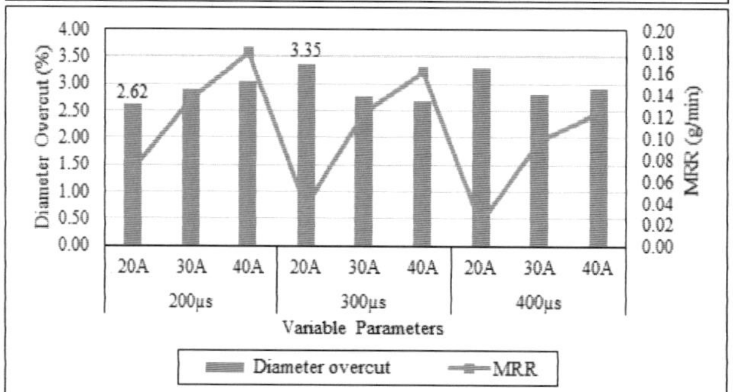

c) Diameter overcut against MRR for third hole

Figure 6. Effect of peak current and pulse duration on diameter overcut against MRR by repetition used of electrode.

The effect of repetition used electrode on the machining accuracy of Inconel 718 also more profound. Based on Figure 6, at low peak current 20A shows that vary on machining accuracy by repetition used of electrode (first, second and third). However, by increasing the peak current the machining accuracy indicate to consistent on repetition used of electrode. This is due to the suspension of black layer on the surface of electrode that cause the erosion of material goes fast on the surface of Inconel 718. Therefore, the crater produced were at maximum and leads to constant erosion of materials. This result may assistance towards accurate of machining.

4. Conclusion

This experiment investigates the material removal rate (MRR), surface topography and machining accuracy applying high peak current and pulse duration when EDM machining of Inconel 718. Within selected parameters and experimental condition, the following conclusion could be draw; Peak current appeared to be the main effect to improve the MRR. As the peak current increased, the MRR increased. Then, the effect of pulse duration is inversely proportional with MRR. As increasing the pulse duration increased, the MRR was decrease. The result indicate that the using of the same electrode repeatedly still capable to enhance MRR.

For the case of surface topography, at low MRR resulting the appearance of micro-geometry characteristics where machining damages take placed. This include void, micro-crack, nodules, pork mark and globules. However, the sizes of the nodules and globules increase at high MRR. This resulted increase in surface roughness. It is also believe that, the surface topography conditions also affected by the carbon and material deposited on the electrode surface due to the repetition used of one electrode for the whole experiments.

On the machining accuracy perspective, the diameter overcut shows slightly fluctuated condition but still consistent by increasing peak current. With deposited of black layer on the surface electrode, high MRR is still obtainable and more consistent of machining accuracy. The result shows quite high of machining accuracy even at higher peak current and pulse duration applied.

Acknowledgement
The authors would like to thank to the Universiti Tun Hussein Onn Malaysia and Research Management Center (RMC) for providing the funding of this research under TIER 1 Grant U871.

References

[1] Pawade R S, Joshi S S and Brahmankar P K 2008 Effect of machining parameters and cutting edge geometry on surface integrity of high-speed turned Inconel 718, *Int. J. Mach. Tools Manuf.*, vol. **48**, no. 1, pp. 15–28.
[2] Bharti P S, Maheshwari S and Sharma C 2010 Experimental investigation of Inconel 718 during die- sinking electric discharge machining, *Int. J. Eng. Sci. Technol.*, vol. **2**, no. 11, pp. 6464–6473.
[3] Beri N, Pungotra H and Kumar A 2012 To Study the Effect of Polarity and Current during Electric Discharge Machining of Inconel 718 with CuW Powder Metallurgy Electrode, *Proc. Natl. Conf. Trends Adv. Mech. Eng.*, pp. 476–481.
[4] Ho K H and Newman S T 2003 State of the art electrical discharge machining (EDM), *Int. J. Mach. Tools Manuf.*, vol. **43**, no. 13, pp. 1287–1300.
[5] Singh S, Maheshwari S, and Pandey P C 2004 Some investigations into the electric discharge machining of hardened tool steel using different electrode materials, *J. Mater. Process. Technol.*, vol. **149**, no. 1–3, pp. 272–277.
[6] Kuppan P, Rajadurai A and Narayanan S 2008 Influence of EDM process parameters in deep hole drilling of Inconel 718, *Int. J. Adv. Manuf. Technol.*, vol. **38**, no. 1–2, pp. 74–84.
[7] Bharti P S, Maheshwari S and Satyarthi M K 2016 Black Layer Characterization in Electric Discharge Machining of Inconel 718, pp. 214–217.
[8] Rajesha S, Sharma A K, & Kumar P (2012). On electro discharge machining of inconel 718 with hollow tool. *Journal of Materials Engineering and Performance*, **21(6)**, 882–891.
[9] Maradia U, Boccadoro M, Stirnimann J, Beltrami I, Kuster F and Wegener K 2012 Die-sink EDM in meso-micro machining, *Procedia CIRP*, vol. **1**, no. 1, pp. 166–171.
[10] Lee H T and Tai T Y 2003 Relationship between EDM parameters and surface crack formation, *J. Mater. Process. Technol.*, vol. **142**, no. 3, pp. 676–683.
[11] Torres A, Puertas I and Luis C J 2016 EDM machinability and surface roughness analysis of TiB2 using copper electrodes," *Int. J. Adv. Manuf. Technol.*, vol. **84**, no. 9–12, pp. 2671–2688.
[12] Sultan T, Kumar A and Gupta R D 2014, Material Removal Rate, Electrode Wear Rate, and Surface Roughness Evaluation in Die Sinking EDM with Hollow Tool through Response Surface Methodology, *Int. J. Manuf. Eng.* pp. 1–16.

[13] Ghewade G and Nipanikar S 2011, Experimental study of Electro Discharge machining for

Inconel material, *J. Eng. Res. Stud.*, vol. **II**, pp. 107–112.

IOP Conf. Series: Materials Science and Engineering **607** (2019) 012007 doi:10.1088/1757-899X/607/1/012007

Microbial study of pH sensitive starch based film using agar diffusion method (zone inhibition assay)

Nozieana Khairuddin[1*], Ida Idayu Muhamad[2], Wan Aizan Wan Abdul Rahman[2] and Bazlul Mobin Siddique[3]

[1*]Department of Engineering, Faculty of Agriculture, Science and Technology, Universiti Putra Malaysia Bintulu Sarawak Campus, 97008 Bintulu Sarawak, Malaysia
[2]Department of Bioprocess and Polymer Engineering, School of Chemical and Energy Engineering, Faculty of Engineering, Universiti Teknologi Malaysia, 81310 Skudai, Johor Bahru Malaysia

[3]School of Engineering, Swinburne University of Technology Sarawak Campus, 93350 Kuching Sarawak, MalaysiaJohor Bahru, Malaysia

*Email corresponding author: nozieana@upm.edu.my

Abstract. Active and smart packaging is a promising form of food packaging that offers a great economical potential due to consumer demand for a packaging that accommodate a hectic way of life. An antimicrobial film with pH colour indicator (pHF) can be made by incorporating suitable antimicrobial (AM) agent and colour indicator into food package matrices whilst applying a bio switch concept to inhibit the pathogenic microorganisms and respond automatically to changes (external stimuli) in the environment. The present work aimed to study the developed formulation of hydroxyethylcellulose (HEC)/wheat-starch based pHF film in which the active compound, thymol (0.5, 1, 1.5, 2, and 2.5% w/w) and 50:50% w/w bromothymol blue and methyl red (as the colour indicator) against microbial growth. A solution casting method was used in the film preparation while thymol and colourant were incorporated prior to casting. The effect of thymol showed a range of microbial inhibition zones of 16.3 - 26.4% and 22.1 - 39.9% towards E. coli and B. subtilis, respectively. Whilst, a lower inhibition zone of 0.4 - 5.1% was demonstrated for fungus A. niger.

Keywords: Active and smart packaging; antimicrobial; thymol; hydroxyethylcellulose; methyl red; bromothymol blue

1. Introduction

The food and beverage packaging has dramatically shifted from traditional to advanced packaging. Traditional packaging only addresses issues related to protection from external factors. However, advanced packaging interacts internally (active packaging) and externally (intelligent packaging) with the environment and enhances the visual appeal of the products. Therefore, manufacturers of food and beverages and packaging manufacturers are looking into the way the physical, chemical and microbial textures change inside the packaged food. The requirement over smart or intelligent packaging is changing rapidly due to the world awareness on environment. The concerns are not only on the materials and productions, but also on legal side related to the environmental concerns.

The general concept of bio-switch describes a system with capability to detect and respond automatically to changes (external stimuli) in the environment. For instance, the external stimulus may be a change in pH, or the presence of certain metabolites from biological activity. The bio-switch converts this stimulus into a particular functionality. Materials with ability to entrap compounds with a specific function which are released on an external stimulus from the environment are created. In order to be used in active packaging, biopolymer-based particles containing antimicrobial will only be released in the case of initial microbial contamination. The bio-switch particles monitor the releasing system by the stimulus of a microbial contamination that actively add or emit compounds i.e. antimicrobials, antioxidants, and preservatives to the packaged food or onto the surface of the package [1].

The current work is a novel combination of active packaging technology and smart packaging concept where the stimulus of a microbial contamination is further incorporated with an indicator to signal the conditions in the packaged food. The present work aimed to study the developed formulation of hydroxyethylcellulose (HEC)/wheat-starch based pHF film in which the active compound, thymol and the duo-colour indicator against microbial growth.

2. Materials and Method

2.1. Materials

The main based of the films which were wheat starch $(C_6H_{10}O_5)n$ and hydroxyethyl cellulose were supplied by Merck. Thymol (C10H14O) was purchased from Sigma-Aldrich (Malaysia), methyl red (MR), bromothymol blue (BB) were purchased from Fluka, Glyoxal was purchased from Merck and glycerol was purchased from HmbG chemicals.

2.2. Film Preparation

In this research, 0.5 g thymol was dissolved in 20 ml of absolute ethanol. Then 0.01 g of bromothymol blue and methyl red was added to the solution respectively and stirred well. The solution was then filtered using filter paper. The filtrate was added to the 80 ml distilled water containing 4 g of HEC and 5 g of wheat starch. After the solution was completely dissolved, 5 ml of glycerol (HmbG Chemicals) and 5 ml of glyoxal was added and the mixture was heated slowly to a mild boiling. Films were casted into square plate (20 x 20 cm). The casting plate was placed for 24 h in an oven (Memmert) set at 60°C. The same step was repeated for the preparation of 1, 1.5, 2 and 2.5% (w/v) of thymol. The control film was prepared with no thymol being aded to the film solution.

2.3. Microbiological Study of using Agar Diffusion Method (Zone Inhibition Assay)

The present sub-topic discussed the method to determine the efficacy of thymol incorporated into starch-based film to inhibit the growth of microorganism. The antimicrobial activity testing was based on the agar diffusion method.

The strain selection represented typical spoilage organism groups commonly occurring in various kinds of food products. The strains were as follows: (1) *Escherichia coli*, a conventional hygiene indicator organism, a Gram-negative rod belonging to the same family of *Enterobacteriaceae* as for example *Salmonella*. (2) *Bacillus subtilis*, a Gram-positive rod capable of forming heat-resistant spores. Spores and vegetative cells of *Bacillus* species are widely distributed in nature and are common for example in cereals. For the agar plate test, the starch-based films containing AM agent were cut into six squares (0.5 cm x 0.5 cm). Six sample squares were then placed onto the plate which was spreaded with bacteria (0.1 mL per plate). The same tests were performed using other film containing, stated: thymol in various concentrations. Duplicate agar plates were prepared for each type of film and control film. The agar plates were incubated at 37°C for 48 hours.

3. Results and Discussion

3.1 Film Appearances

Figure 1 shows the visual aspect of all formulations of pH sensitive films incorporated with 0, 0.5, 1, 1.5, 2, and 2.5 % w/v of thymol for film pHF-0, pHF-0.5, pHF-1, pHF-1.5, pHF-2, and pHF-2.5 respectively with the addition of colour indicator. The control film produced without thymol being added was completely translucent, smooth, and glossy (pHF-0). No visual changes were observed in the films containing low concentrations of thymol, compared with the control film; they were still translucent and clear.

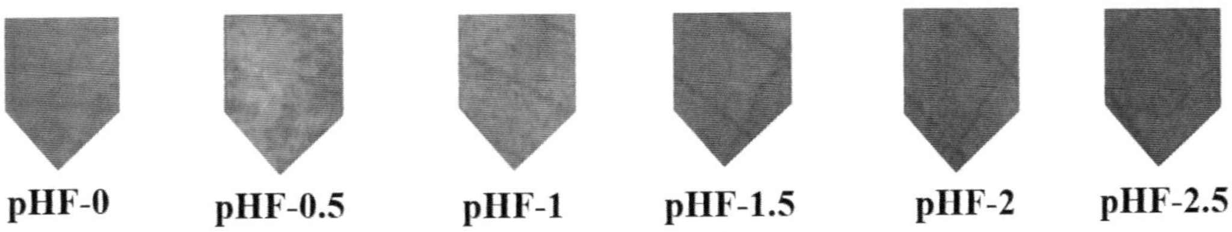

Figure 1. Visual appearance of the films with different concentration of thymol

Figure 2 shows the visual colour changes of the prepared indicator solutions. A reliable colour changes are required for production of indicator label as it will be a suitable communication-tool for the consumers [2]. The main purpose for applying colorimetric mixed-dye-based indicators to food packaging is to easily and reliably monitor the level of food spoilage of packaged food products in a non-destructive manner during distribution and retail sale [3]. It was found that indicator solutions showed a clear spectrum from bright light orange to dark blue when exposed to a pH level range of 1 - 14. However, similar colour spectrum could be monitored with naked eyes for indicator solutions in the range of pH 1 till 5. All the solutions changed into bright light orange, red-orange and orange.

An obvious transformation of the colour indicator was observed in the range of pH 5 till pH 8 (Figure 2). It was found that indicator solutions showed a clear spectrum from bright orange, red orange, yellow, green, and blue when exposed to pH 5, 6, 7, and 8 respectively. This is not surprising as the colour indicator of MR best works in the range of pH 4.4 to 6.2, while BB best works in the pH ranges from 6 to 7.6 [4]. The most obvious finding emerged from the analysis is that the most obvious colour change of the indicator solutions found in this study falls in the range of most pH for food deterioration. Moreover, it is also in the range of optimum pH for microbial growth which was the main factor of food spoilage [5].

Whereas, a similar blue spectrum could be observed with unaided eye for all indicator solutions in the pH ranges of 9 to 14. This is due to the effect of BB which changed to blue when reacted with pH more than 7.6. These findings have significant implications for the understanding of how the indicator solution will work in the HEC-wheat starch based film incorporated with thymol.

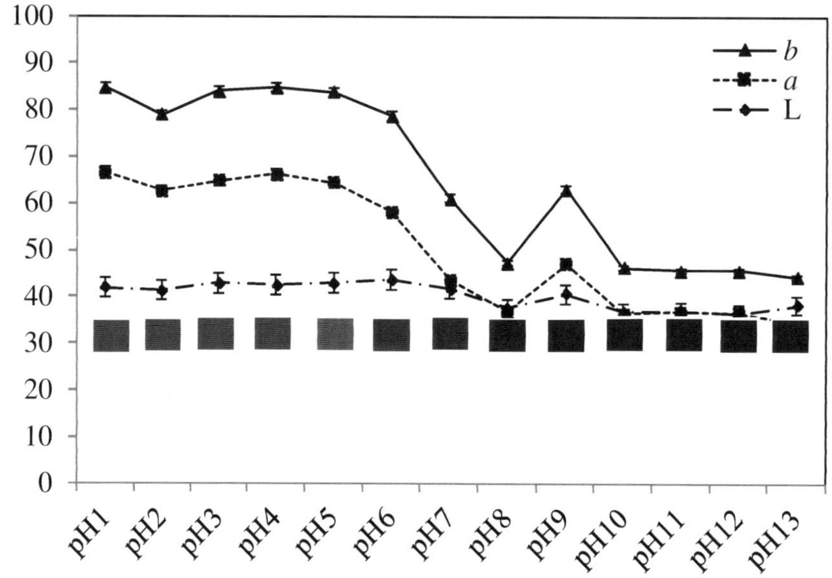

Figure 2. Change in colour of indicator solutions in response to pH buffer

3.2 Microbiological assessment of pHF film using agar diffusion method

The present test focus on the efficacy of the films to inhibit the growth of selected Gram positive food pathogenic bacteria i.e. Bacillus subtilis and Gram negative food pathogenic bacteria i.e. Escherichia coli. Agar plate test, also known as zone inhibition assay was performed as a preliminary step to screen the antibacterial activity of all films formulations, in an effort to select film formulations with high antibacterial activity against the test bacteria. The inhibitory zone in agar diffusion test can be affected by the solubility and diffusion rate of the test compounds in agar medium, thus agar diffusion test does not accurately reflect the antimicrobial effectiveness of the test compounds [6, 7].

All samples were examined for possible inhibition zones after incubation at 37°C for 24 hours. Table 2 lists the calculated inhibition area for each plate test. A similar finding to the previous research had been acquired. The control films showed no inhibition area and colonies were formed all over the plate for both types of microorganisms: the gram positive and gram negative bacteria. It can therefore be assumed that the colour indicator (MR and BB) did not have any antibacterial properties. Hence, the antibacterial activity provided by the films is primarily derived from thymol.

Table 1. Analysis of the zone of inhibition data in agar plate test for *E. coli, B. subtilis* and *A.niger* at 37°C in the presence of pH sensitive film incorporated with thymol

Films	Concentration of thymol in each film (w/v)	Diameters of growth inhibition zone (mm ± SD)		
		E. coli	*B. subtilis*	*A. niger*
		48 h	48 h	48 h
pHF-0	0.0%	nd	nd	nd
pHF-0.5	0.5%	$41.0 \pm 0.20^{a,b}$	44.3 ± 0.31^{b}	22 ± 0.00^{d}
pHF-1	1.0%	39.0 ± 0.00^{a}	$42.3 \pm 0.21^{a,b}$	$19.5 \pm 0.07^{d,e}$
pHF-1.5	1.5%	39.3 ± 0.11^{a}	$42.3 \pm 0.06^{a,b}$	$19.5 \pm 0.35^{d,e}$
pHF-2	2.0%	38.0 ± 0.00^{a}	45.0 ± 0.85^{c}	17.5 ± 0.07^{f}

pHF-2.5	2.5%	36.3 ± 0.06^{a}	45.3 ± 0.32^{b}	6.5 ± 0.07^{g}

nd: no detection of inhibition zones.
*means in the same column with the same letter are not significantly different (P>0.05)

Clear zones on the agar indicating the inhibitory action of thymol against the growth of bacteria tested were observed around films incorporated with thymol as AM agents (Figure 3). pH sensitive films containing thymol effectively inhibit the growth of E. coli and B. subtilis. However, no significant effect on the concentration of thymol caused the inhibition effect. Yet again, as discussed in previous research by Tiwari and friends in 2009, thymol has the ability to kill the microorganism in wide range of concentration including a low concentration (0.5 % w/v) as chosen in this study [8].

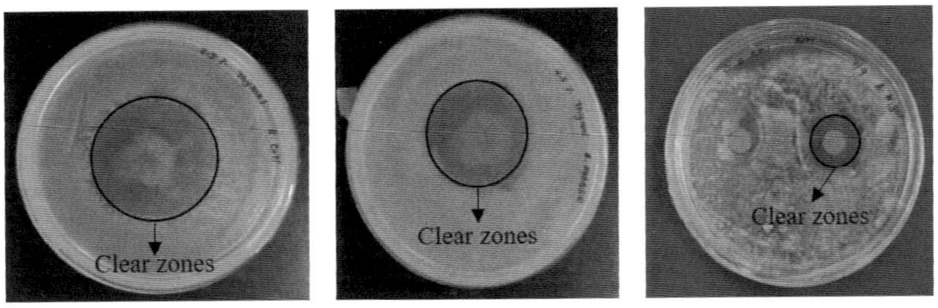

E. coli *B. subtilis* *A. niger*

Figure 3. Inhibition of *Escherichia coli*, *Bacillus subtilis* and *Aspergillus niger* on solid media by pHF film incorporated with thymol after incubation for 24 hours at 37°C.

Conclusion

The present paper discussed the antibacterial properties of the pHF films incorporated with thymol in the range of 0.5 % to 2.5 % (w/v). As supported by previous study, essential oils have a synergistic effect towards inhibiting the microbial growth in a wide range of concentration even in a small amount. In conclusion, pH colour indicators give no significant effect on the ability of thymol to inhibit the microbial growth compared to pHF-0 film. It is not surprising to perceive an effective inhibition of the growth of selected gram positive (*B. subtilis*) and gram negative (*E. coli*) bacteria in wide range of studied concentration of thymol. pHF films containing thymol has also proven to have the ability to inhibit the growth of *A. niger*.

Acknowledgement

The authors would like to thank the Ministry of Education for FRGS grant (FRGS/1/2018/TK05/UPM/02/8) and Research Management Centre UPM for their support of this study.

References

[1] Muhamad II, Salleh E and Khairuddin N 2008 Active and smart Packaging using a bio-switch concept. In Lee CT and Muhamad II. Chapter in Special Topic in Bioprocess Engineering Volume **II**.
[2] Nopwinyuwong A, Boonsupthip W, Pechyen C and Suppakul P 2012 Preparation of polydiaceylene vesicle and amphiphilic polymer as time-temperature indicator. *Adv Mat Res* **506** 552–5.

[3] Nopwinyuwong A, Trevanich S and Suppakul P 2010 Development of a Novel Colorimetric Indicator Label for Monitoring Freshness of Intermediate-moisture Dessert Spoilage. *Talanta* **81** 1126 – 1132.
[4] Sabnis, R. W 2018 Handbook of Acid-Base Indictors. CRC Press, Boca Raton.
[5] Seema R 2015 Food spoilage: Microorganism and their prevention. *Asian J Plant Sci Res* **5(4)** 47 – 56.
[6] Kim D H, Na S K and Park J S 1995 Preparation and Characterization of Modified Starch-based Plastic Film Reinforced with Short Pulp Fiber. II. Mechanical Properties. *J Appl Polym Sci* **88** 2108-2117.
[7] Han J H 2003 Antimicrobial Food Packaging. In: Ahvenainen R. *Novel Food Packaging Techniques*. England: Woodhead Publishing Limited. 50-95.
[8] Tiwari B K, Valdramidis V P, O'Donnel C P Muthukumarappan K, Bourke P and Cullen PJ. 2009 Applications of Natural Antimicrobials for Food Preservation. *J Agric Food Chem* **57**, 5987 – 6000.

IOP Conf. Series: Materials Science and Engineering **607** (2019) 012008 doi:10.1088/1757-899X/607/1/012008

Microstructure mechanism of the hydroxyapatite densification process

M A Selimin[1*], T C Lee[1], Haq R H A[2], M I Idris[2] and H Z Abdullah[2]

[1] Department of Production and Operation Management, Faculty of Technology Management and Business, Universiti Tun Hussein Onn Malaysia, 86400 Parit Raja, Batu Pahat, Johor, Malaysia

[2] Department of Manufacturing Engineering, Faculty of Mechanical and Manufacturing Engineering, Universiti Tun Hussein Onn Malaysia, 86400 Parit Raja, Batu Pahat, Johor, Malaysia

* Corresponding author: aliselimin@uthm.edu.my

Abstract. Hydroxyapatite (HA) is a well-known materials used as biomaterials in various applications. Numerous studied had been done for unlocking the pros and cons of this material. Sintering process has been used for extraction of pure HA that have similar Ca/P ratio to human bone mineral. This study aims to investigate the effect of sintering temperature on the properties of synthetic HA as well as to develop the mechanism of this densification process. Synthetic HA powder (Sigma-Aldrich #04238) was mixed together with binders, namely polyvinyl alcohol (PVA) and polyethene glycol (PEG) in wet mixing condition using ball mill. Pressed HA (15 g, 65 mm x 12 mm) was sintered with temperature ranging from 1000°C to 1200°C with heating rate and cooling rate of 10°C/min. It has been found that, pure HA was obtained when the HA was sintered with temperature ranging from 1000°C to 1200°C. The thermogravimetric analysis showed that there are there were two mass loss stages on the heating process from room temperature to 1250°C; 100-500°C (dehydroxylation) and 680-1080°C (decomposition). From the microstructure data, the mechanism of the densification process of HA was developed.

1. Introduction

Hydroxyapatite (HA) is a bioceramic material which is chemically and mineralogically similar to the non-organic component of natural bone and teeth (highly biocompatible and osteoconductive). It is a well-known material for treating bone defects. Current biomedical applications of bioceramics are commonly carried out in orthopaedic and dental implant coatings, dental implant materials, bone-graft substitute materials, and bone cement [1].

HA has no carcinogenic properties and does not trigger allergic reactions. Nevertheless, despite its advantages, the application of this bioceramic material is limited due to its other less desirable characteristics such as poor resorption, a substantially high Young modulus, and low fracture toughness [2-3].

HA had been studied for biomedical application due to its similarity in term of chemical and structure with natural bone and teeth [4-5]. HA is also used as biocompatible coating material to improve the bonding quality between metallic substrates and the surrounding bone tissue following implantation [6].

One of the factors that limits the consumption of this material is the brittle property of calcium phosphate group.

The beneficial biocompatible properties of HA are well documented in previous studies. It can be integrated into the human body, and at the same time, the body is not wise as to the invasion of a foreign body, although it is only a friendly invasion. The most interesting property is the bonding of HA to the bone to form indistinguishable unions. However, poor mechanical properties (in particular fatigue properties) mean that this material cannot be used in bulk form for load bearing applications such as orthopaedic [7-9]. It is important to consider the properties, morphology and porosity of HA, especially after the sintering process before considering it to be used as a biomedical material in the medical industry. The HA had a Ca/P ratio 1.67 which is one of the important characteristics of bioactive material as it can provide more sites for bone growth in the human body [10].

There is the existence of countless number of phosphate compounds, as well as highly complex system of calcium phosphate system. This is further complicated by the sensitive stability of phosphates to the minor changes in terms of composition, pH, and reaction conditions (e.g. temperature). It is important to acknowledge the fact that the purity and particle characteristics of the final synthesised powder can affect the bioactivity, mechanical and biological dissolution properties [11]. These characteristics could determine the medical application of the material, which makes it imperative in developing a synthesis method that enables the control of crystal morphology, chemical composition, crystallinity, particle size distribution, and agglomeration.

Thermal stability of HA is an important issue since the implants are usually prepared by high sintering temperature (1000-1350ºC) of compacted HA powder. The temperature and time of thermal treatment are the most important parameters that could influence the HAP properties. During the sintering process, there is the occurrence of two separate processes that affect the morphology of solids: densification and grain growth [12]. Two of them are thermally activated, but only densification process is desired. In the first step of sintering process (in temperature below 750ºC), the internal powder expansion is observed. Subsequently, the initiation of higher temperature compaction process would lead to the removal of pores from the material. Finally, the sample is shrinked and reached its maximum density in sintered form. Sintering is conducted to achieve the highest density while reducing grain growth [13].

Decomposition of HA at high temperatures is frequently encountered during the fabrication of HA ceramics. This process inhibits the sinterability of the apatite ceramics and completely transforms HA into a mixture of TTCP and TCP, and α-TCP can readily react with water. When the reactions occur in the bioceramics under physiological conditions, the mechanical integrity of HA would deteriorate. The microstructures could correlate well with the density (directly proportional), and the bulk density of HA increases when increasing the sintering temperature [14]. Increasing the sintering temperature beyond 1050ºC could resulted in grain growth, and almost full densification was accomplished when compact HA was sintered at >1050ºC. The present study deals with the extraction of pure HA via sintering process with different mixing ratio with binders. The mechanism of HA densification process is developed in this study.

2. Materials and Methods

2.1. Preparation of sintered HA

The HA powder (Sigma-Aldrich, #04238) was used to produce a rectangular specimen to study the properties of these materials with varied processing parameters. The rectangular HA specimens were prepared by mixing binders, namely polyvinyl alcohol (PVA) and polyethene glycol 6000 (PEG) in wet mixing condition using ball mill. This process took 24 hours and 160 rpm rotational speed. Then, the slurry was dried in an oven. Next, the slurry was crushed and sieve through a 250 μm mesh. The HA powder was then pressed using uniaxial press machine to shape it into a rectangular (65 mm x 12 mm). After that, the green body of HA specimen undergo a sintering process with temperature ranging from 1000ºC to 1200ºC. This was accompanied with heating and cooling rate of 10ºC/min. The mixing ratio of the HA specimen are tabulated in Table 1. The sintering temperature was set based on the thermogravimetric analysis for HA, PEG and PVA.

Table 1. HA specimen mixing ratio and its abbreviation.

Abbreviation	Material (% wt)		
	HA	PEG	PVA
HAP 1	100	4	0
HAP 2	100	3	1
HAP 3	100	2	2
HAP 4	100	1	3
HAP 5	100	0	4

2.2. Characterisation and testing

The thermogravimetric analysis (Linseis, Thermobalance Simultaneous Thermal Analysis (STA) TG + DTA, Germany) was used to identify the mass loss of the HA, PEG and PVA. The mineralogical properties of HA were characterised using X-ray diffractometer (XRD, Bruker D8 Advance). The microstructure of failure surface was captured using field emission scanning electron microscopy (FESEM, JSM-7600F, Joel Ltd., Japan).

3. Results and Discussion

3.1. Thermogravimetric analysis (TGA)

The initial weight of HA, PEG and PVA was set at 27.0 mg with heating rate 10°C/min. The TG curve showed that there were two mass loss stages on the heating process from room temperature to 1250ºC; 100-500ºC and 680-1080ºC. At the first stage, removal of adsorbed water from the surfaces and pores of hydroxyapatite took place. Meanwhile, the dehydroxylation of the decomposed HA powder occurred on the second stage at the temperature of 680-1080ºC. The total mass loss at Stage 1 and Stage 2 were 5.5 % and 1.2 % respectively. According to previous studies, sintered HA at 1250ºC and above (Stage 3 in Figure 1) will lead to the formation of other phases such as alpha tricalcium phosphate (α-TCP) and other resorbable phases of calcium phosphates [15]. These phases had poor mechanical property and higher solubility than pure HA phase. In addition, the amount of HA mass loss increased with the increment of sintering temperature. The results were parallel with previous studies using another commercial HA powder [16].

Moreover, it was expected that hydroxyapatite powder with pure HA phase to be the final product of HA. Therefore, to achieve this stage, the sintering temperature was ranged between 1000-1200ºC. At these temperatures, the binder agents burned out and increased the density and mechanical properties of the final product [17]. It was observed that total mass loss (burnt off) for PEG 6000 was occurred at 600ºC and above 700ºC for PVA. In this study, binders were used to hold the HA compacted particles in the pressing process to form rectangular specimens. Both binders did not give any effect to HA because it burnt off at high sintering temperature (> 700ºC).

3.2. Mineralogical

Figure 2(a) indicated that the HA phase was stable during sintering up to 1200ºC and there was no presence of TCP (α-TCP or β-TCP). It was observed that long soaking or sintering time did not lead to any changes in phase groups. On top of that, there was no glassy phase formation detected in any specimens, which may cause the material to be brittleness [18].

Furthermore, the XRD diffractions of HAP 3 specimens as a function of sintering temperature are presented in Figure 2(b). The diffractions demonstrated that there were no other phases except HA peaks on the specimens, even when the sintering temperature were varied up to 1200ºC. This was in

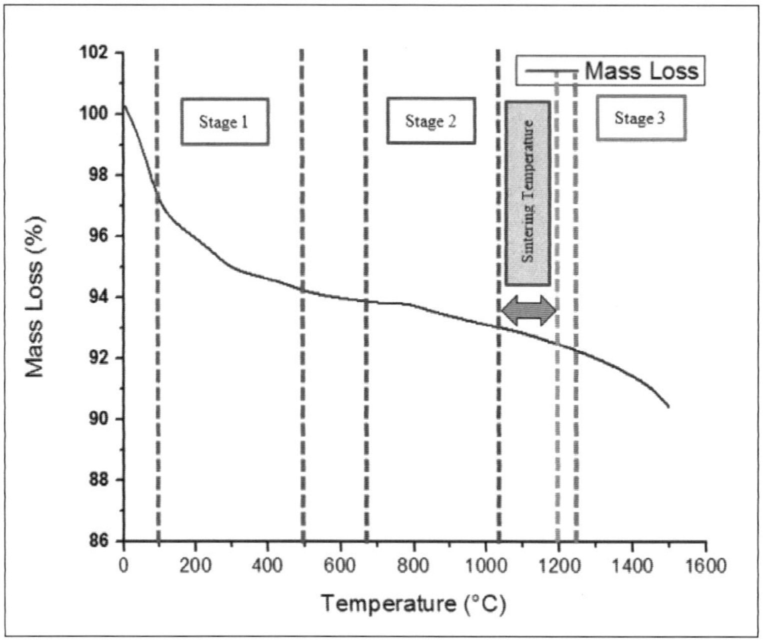

Figure 1. TG curve of HA powder at 10ºC/min heating rate.

agreement with previous findings, which was that only HA phases of calcium phosphate groups will be present on the specimen unless higher sintering temperature is applied (> 1250ºC) [19]. Moreover, other diffractions also demonstrated similar pattern (peak intensity). This showed that sintering temperature in the range of 1000ºC to 1200ºC did not have much influence on its peak intensities. Since binders (PEG and PVA) were burnt off completely at sintering temperature between 600ºC - 700ºC, there was no PEG or PVA peaks observed (this was confirmed with the TGA results shown in Section 3.1). The binders provided better compaction during HA sample preparation (pressing process).

The XRD profiles of sintered HA as a function of binder ratio are shown in Figure 6.14 there was no phase decomposition of HA into secondary phases such as α-TCP, β-TCP, TTCP or CaO. Also, the addition of binders (PEG and PVA) onto HA specimen with varied mixing ratios from 1 wt % to 4 wt % did not influence the XRD phase of HA. Moreover, there was no remarkable PEG and PVA peaks observed on the diffractive angle of 23.34°, as marked on the figure where it was supposed to be present [20]. Other than that, all diffractions showed similar patterns, as discussed earlier. All peaks and intensities were also identical.

3.3. Microstructure of failure surface
Figure 3 indicated that the number of open pores were reduced by increasing the sintering temperature, enabling the increase of the HA specimen grain size and eliminate its porosity. In Figure 3(a), numerous small open pores on the cross-section of failure surface of HA at low sintering temperature (1000ºC). Then, the number of open pores decreased as the sintering temperature increased. At low temperature, the specimen exhibited many small sizes of open pores. The failure began at pores, passes through the grains and went further into another pores. Unlike at low temperature, high sintering temperature led to the failure beginning at in between two grains, before it moved on to the nearest pores. This was due to the reduction in the number of pores because the increased in grains size [21]. At high temperature, the small pores merged and formed less and bigger pores.

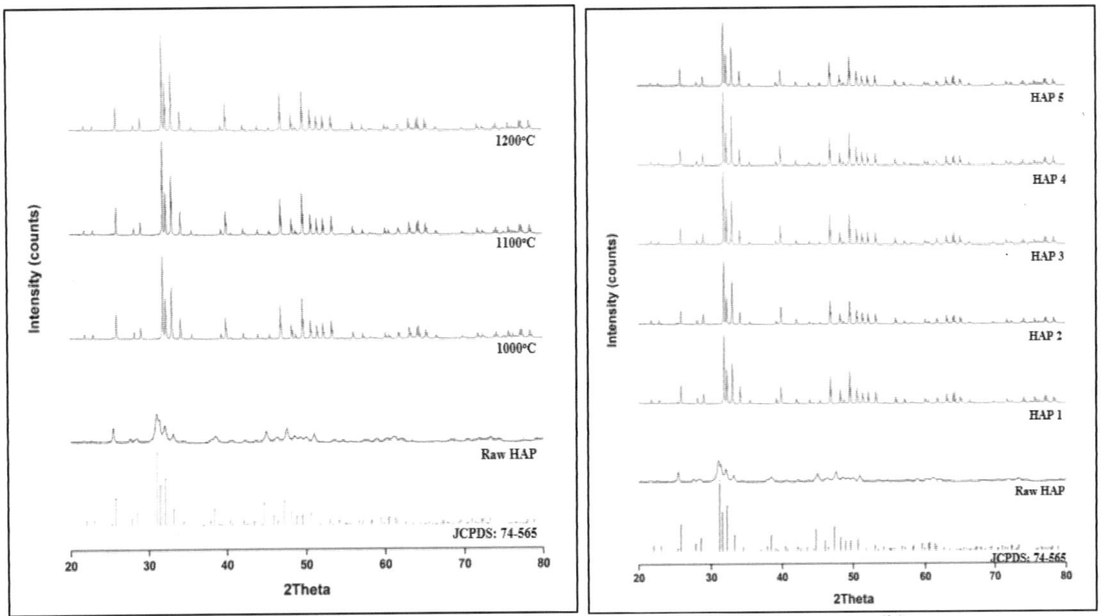

(a) (b)

Figure 2. XRD patterns: (a) sintered HAP 3 as a function of sintering temperature, (b) sintered HA at 1200°C as a function of binder ratio.

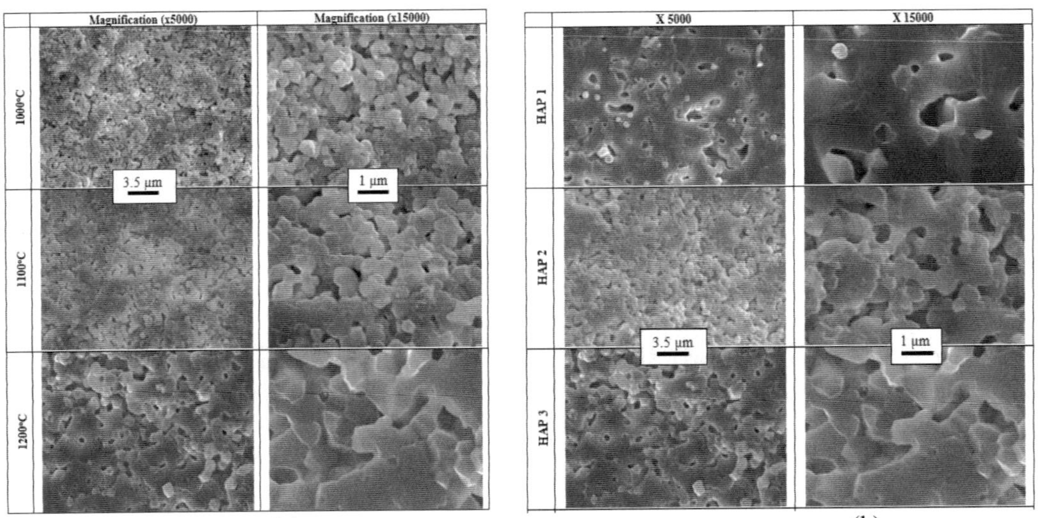

(a) (b)

Figure 3. FESEM micrograph: (a) sintered HAP 3 as a function of sintering temperature, (b) sintered HA at 1200°C as a function of binder ratio.

The microstructure images of HA specimens at 1200°C as a function of binder mixing ratios are presented in Figure 3(b). There seemed to be no major difference in the properties of HA specimens at the same sintering temperature, except for HAP 2. The FESEM micrographs supported the mechanical properties observed throughout the study, in which the morphology for all HA specimens at same temperature had similar microstructure except for HAP 2. HAP 2 had slightly different microstructure in term of grain size and numbers of pores that made its mechanical properties different to other specimens.

Moreover, although binders gave small effects on the properties of HA specimens, it still influenced the microstructure of the failure surface. This can be seen on the specimen with the highest PVA element (HAP 5), which had many large pores. This was caused by the PVA not providing better compaction during the compression process, leading to the presence of more pores. Consequently, the mechanical strength and density of the specimen will lower.

3.4. Mechanism of HA densification process

The mechanism of HA densification process is developed based on the microstructure analysis done in **Section 3.3**. Based on the micrographs in Figure 3, it can be found that there are five stages of densification (grain growth) of HA specimen (Figure 4). At room temperature, green body of compacted HA have brittle properties due to highest porosity (coloured with green) and lowest density (density is directly proportional with porosity). As the temperature increasing, the size of grains is increased which cause to contact between grains being oriented, developed and grown. The pores healing happen (size and number of pores reduce) as the temperature increased that lead to excellent mechanical properties.

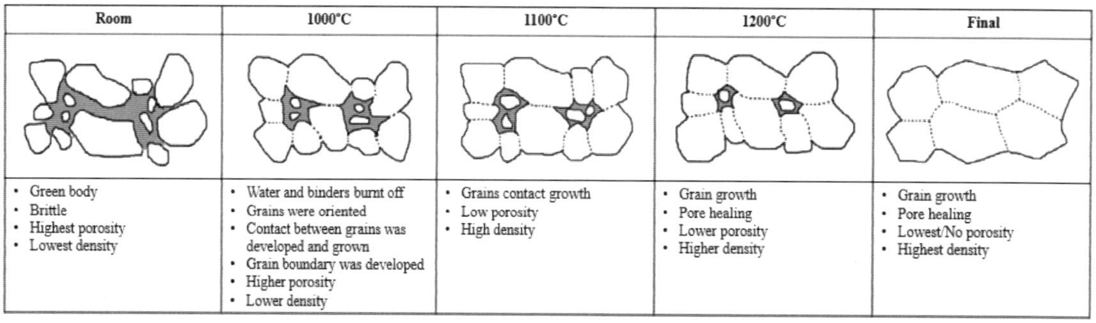

Figure 4. Microstructure mechanism of grains growth (densification) and it explanations.

4. Conclusion

There are two stages of grains growth process for HA; Stage 1 (removal of absorbed water) at 100-500°C and Stage 2 (dehydroxylation) at 680-1080°C. The densification process started with elimination of impurities and adsorbed water (hydroxyl groups) from the surface and pores. As the sintering temperature increasing, the dehydroxylation take place and this process leads to reduction of size and number of pores until it totally removes from the HA at Stage 2 of HA densification process. Reducing in number of pores (porosity) enables the HA gain better mechanical properties (overcome the brittleness of HA) and this widen the application of HA from non-load bearing application to load bearing application. In addition, the size and boundaries of grain increased with increment of temperature and this provides better mechanical properties in term of strength and many more. There are five microstructure mechanism of grains growth for HA developed in this study based on the data from the FESEM analysis.

Acknowledgment

The authors gratefully acknowledge Universiti Tun Hussein Onn Malaysia and Ministry of Education Malaysia for the financial support via research grant FRGS K105, GPPS U978, and MPC K131.

References

[1] Borkowska A and Ginalska G 2010. Hydroxyapatite Biomaterials for a Filling of Bone Defects. *Annales Universitatis Mariae Curie-Sklodowska* **23** 45-52
[2] Mohd Pu'ad N A S, Koshy P, Abdullah H Z, Idris M I and Lee T C 2019. Syntheses of

hydroxyapatite from natural sources. *Heliyon* **5** e01588

[3] Best S M, Porter A E, Thian A E and Huang J 2008. Bioceramics: Past, Present and for the Future. *Journal of the European Ceramic Society* **28** 1319-1327

[4] Kanasan N, Adzila S, Mustaffa N A, Rahman M N A, Hassan M F, and Panerselvan G 2018. Effect of sintering on hydroxyapatite/sodium alginate properties. *International Journal of Integrated Engineering* **10** 78-82

[5] Fara A N K A and Abdullah H Z 2016. Influence of calcination temperature on the microstructure and crystallographic properties of hydroxyapatite from black tilapia fish scale. *Materials Science Forum* **40** 151-155

[6] Chien C S, Han T J, Hong T F, Kuo T Y and Liao T Y 2009. Effects of Different Hydroxyapatite Binders on Morphology, Ca/P Ratio and Hardness of Nd.YAG Laser Clad Coatings. *Materials Transactions* **50** 2852-2857

[7] Sasikumar S and Vijayaraghavan R 2006. Low Temperature Synthesis of Nanocrystalline Hydroxyapatite from Egg Shells by Combustion Method. *Trends in Biomaterials and Artificial Organs* **19** 70-73

[8] Bellucci D, Sola A and Cannillo V 2015. Hydroxyapatite and Tricalcium Phosphate Composites with Bioactive Glass as Second Phase: State of the Art and Current Applications. *Journal of Biomedical Materials Research - Part A* **104** 1030-1056

[9] Chang Q, Chen D L, Ru H Q, Yue X Y, Yu L and Zhang C P 2010. Toughening Mechanisms in Iron-Containing Hydroxyapatite/titanium Composites. *Biomaterials* **31** 1493-1501

[10] Cahyaningrum S E, Herdyastuty N, Devina B and Supangat D 2017. Synthesis and characterization of hydroxyapatite powder by wet precipitation method *IOP Conf. Series: Materials Science and Engineering* **299** 012039

[11] Cox S 2012. Synthesis Method of Hydroxyapatite. *Ceramics* **2** 1-10

[12] Gu Y W, Loh N H, Khor K A, Tor S B and Cheng P 2002. Spark Plasma Sintering of Hydroxyapatite Powders. *Biomaterials* **23** 37-43

[13] Champion E 2013. Sintering of Calcium Phosphate Bioceramic. *Acta Biomatertals* **9** 5855-5875

[14] Ramesh S, Tan C Y, Bhaduri S B, Teng W D and Sopyan I 2008. Densification Behaviour of Nanocrystalline Hydroxyapatite Bioceramics. *Journal of Materials Processing Technology* **206** 221-230

[15] Nath S, Biswas K, Wang K, Bordia R K and Basu B 2010. Sintering, Phase Stability, and Properties of Calcium Phosphate-mullite Composites. *Journal of the American Ceramic Society* **93** 1639-1649

[16] Wang T, Dorner-Reisel A and Muller E 2004. Thermogravimetric and thermokinetic investigation of the dehydroxylation of a hydroxyapatite powder. *Journal of the European Ceramic Society* **24** 693-698

[17] Temenoff J S and Mikos A G 2008. Biomaterials: The Intersection of Biology and Materials Science. *Pearson/Prentice Hall*

[18] Karacayli U, Gunduz O, Salman S, Ozyegin L S, Agathopoulos S and Oktar F N 2009. Effect of Sintering Temperature on Mechanical Properties and Microstructure of Sheep-Bone Derived Hydroxyapatite (SHA). *IFMBE Proceedings* **23** 1271-1274

[19] Monmaturapoj N and Yatongchai C 2010. Effect of Sintering on Microstructure and Properties of Hydroxyapatite Produced by Different Synthesizing Methods. *Journal of Metals, Materials and Minerals* **20** 53-61

[20] Ahmad M, Tay M Y, Shameli K, Hussein M Z and Lim J J 2011. Green Synthesis and Characterization of Silver/Chitosan/Polyethylene Glycol Nanocomposites without any Reducing Agent. *International Journal of Molecular Sciences* **12** 4872-4884

[21] Abdullah H Z 2002. *Kesan Zink Oksida (ZnO) Ke Atas Sifat Seramik Hidroksiapatit. Master Thesis. Universiti Sains Malaysia*

IOP Conf. Series: Materials Science and Engineering **607** (2019) 012009 doi:10.1088/1757-899X/607/1/012009

Musculoskeletal symptoms and ergonomic risk assessment among production operators at manufacturing industries: A review

Ganaswaran Segar and Mohd Nasrull Abdol Rahman

Department of Manufacturing and Industrial Engineering, Faculty of Mechanical Engineering and Manufacturing, Universiti Tun Hussein Onn Malaysia (UTHM), 86400 Batu Pahat, Johor, Malaysia

Corresponding author: mnasrull@uthm.edu.my

Abstract. Musculoskeletal symptoms are prevalent and represent the most common health problem among manufacturing operators in industrial-developing countries, with considerable costs and impact on quality of life. Because of incorrect workstations and lack of employee education in basic biomechanical principles, many of the workers are still in high risk injury. The aim of this study is to understand the impact of ergonomics implementation among production operators as the risk factors of musculoskeletal symptoms. Operators are normally exposed to a variety of risk factors such as awkward and static postures, repeat movement, and powerful effort. Manufacturing operators are thus considered at risk from developing musculoskeletal symptoms (MSSs). This review evaluates selected papers in manufacturing industries that have studied risk factors of musculoskeletal symptoms among manufacturing operators. Furthermore, other related industry studies have been reviewed as applicable. To understand the risk factors of musculoskeletal symptoms among manufacturing operators, it is recommended that future studies be required to assess these risk factors among manufacturing operators.

1. Introduction

In a working population, musculoskeletal symptoms (MSSs) are the most common and leading cause of occupational injury and disability in developed and industrial developing industrial countries [1][2]. Musculoskeletal symptoms (MSSs) in a working population are the most common and leading cause of occupational injury and disability in developed and industrial developing countries [1][2]. Production operators are potentially exposed to ergonomic hazards or risk in the manufacturing process. Age, gender, height, weight, smoking status and physical activity may influence the occurrence of MSSs [4].

Lower back pain (LBP) and shoulder pain were regularly described as a high risk to manufacturing operators in most manufacturing industries where there was exposure to different musculoskeletal symptoms and mostly uncomfortable postures and repetitive movement. [5] [6] [7]. Studies claimed this disorder was caused by continuous standing, reaching, bending, pushing, pulling, and twisting among manufacturing operators. In addition, the worker's body's shoulder pain is usually exposed to physical load in different tasks with the risk of developing MSSs [8].

2. Risk Factors among Production Operators in Manufacturing Industries

In almost every year, musculoskeletal symptoms among manufacturing operators are increasing. In the absence of an efficient work injury prevention program and the execution of incorrect methods, "poor technique" is widely reported as causing injury among manufacturing operators [9]. According to Social Security Organization, SOCSO (2016), it was reported that many workers from all around industries are experiencing many types of musculoskeletal symptoms (MSSs) and the cases are still increasing [10]. In recent years, the study of risk factors affecting workplace musculoskeletal health issues has been an important area of research focus. These risk factors were commonly grouped as individual, physical, and psychosocial or organizational factors in the work environment [11]. Further research findings can be found in Table 1 from several reviews of risk factors of musculoskeletal symptoms among manufacturing operators.

Table 1. Further research findings

AUTHOR	TITLE	FINDINGS
Aghilinejad et al., (2012)	Work-related musculoskeletal complaints among workers of Iranian aluminum industries.	The most frequent musculoskeletal complaints in the neck, shoulder, and lumbar. Study also showed that job duration and age were significantly associated with musculoskeletal symptoms in the different body regions.
Dianat et al., (2015)	Association of individual and work-related risk factors with musculoskeletal symptoms among Iranian sewing machine operators.	Work organization factors including long duration of sitting work without a break and prolonged working hours per shift, as well as other work-related variables such as the number of years worked as an operator and feeling pressure due to work were also found to be associated with the occurrence of musculoskeletal symptoms in different body regions.
Foong MC et al., (2014)	Prevalence of Musculoskeletal Symptoms Among Production Line Workers in a Printing.	The prevalence of musculoskeletal symptoms was 79.6%. The most common musculoskeletal symptoms were from the lower back (48.0%).
Sealetsa, O. J., & Thatcher, A. (2011)	Ergonomics issues among sewing machine operators in the textile manufacturing industry in Botswana	Back, neck and shoulder discomfort are highly prevalent among these sewing machine operators. Prevalence rates of 70%for the upper back, 63%for the lower back, and 61% for the shoulders and mid-back region.
Merisalu et al., (2016)	Predictors and prevalence of musculoskeletal disorders among sewing machine operators	The respondents assessed pain in the neck, lower back, and both on the right and left side of shoulders, elbows, wrists/hands and knees. The questions about individual, physiological and psychological risk factors and health behavior were included.

2.1 Awkward Postures

Posture refers to the position of your body's various parts. Muscles, tendons and ligaments have to work harder and can be stressed in an awkward position [12]. In a case study that conducted manufacturing process automation in Korea. It has been stated that the number of cases of MSSs is increasing frequently in each year. One of the factor related of MSs was awkward postures by workers/operators especially in assembly part section [13]. Operators who work in uncomfortable positions can increase the risk of low back pain [14]. O'Sullivan (2012) also mentioned that Prolonged sitting periods, for example periods exceeding 30 min, are a common aggravating factor for many subjects with low back pain.

2.2 Force of Exertion

Force is the amount of effort our bodies have to make in lifting objects, using tools, or moving. Since, as a rule, there is not enough time for recovery in repetitive work, the more powerful movements develop tiredness much faster. [12]. According to Department of Environmental Health and Safety of IOWA State University (2013), forceful movements refers to the amount of effort needed to perform a task or motion. Younger people have been reported to have a higher predisposition to develop low back MSs, as it is more common for younger people to carry out activities that require strong movements and postures, making them more susceptible to low back MSSs [15]. In contrast, older individuals may perform activities that do not require as much physical effort; they are more likely to perform activities that demand more precise upper limb tasks instead, such as computer-related activities and line assembly tasks [15].

2.3 Repetition Motion

Repetition rate is defined as the average number of movements or exertions performed within a unit of time by a joint or body link or performing similar movements with the same part of the body with little rest or recovery[12]. Repetition puts workers at an increased risk of injury when other risk factors, such as an awkward posture or heavy force, are also present. [16] [17]. According to findings, due to standardized workstations, high incidence of musculoskeletal symptoms and related physical problems in the manufacture of textiles and clothing was highly repetitive. [3]. In some of the earliest studies reported in the ergonomics literature on sewing machine operators, operators reported pain in the past 12 months in at least one region of the body and most often in the lower back, neck and wrist / hand [18] and neck, shoulders and lower limbs. [19].

2.4 Vibration

Vibration results in effects such as damage to body organs caused by high levels of vibration at relatively low frequencies and breakdown of body tissue due to either continuous resonance or high energy vibration absorption. [12]. Exposure to WBV occurs when vibrations are usually transmitted through the feet when standing or when sitting on the legs and hips. WBV, including internal organs, can affect the whole body. The whole body's exposure to vibration (usually through the feet/ buttocks while riding in a vehicle) has some support as a risk of injury. [16] [12]. Van et al. (2016) noted that workplace vibration exposure and work stress were a significant risk factor for neck, shoulder, lower back and upper back discomfort or pain. [20]. Other than that, in a study among operators of construction equipment that carried out by Kittusamy & Buchholz (2004), analyzed that whole body vibration (WBV) and the postural requirements of work (both static and awkward postures) are important risk factors that contribute to the development of musculoskeletal disorders among Operating Engineers (OEs) [21].

3. Musculoskeletal Symptoms (MSSs) among Production Operators at Manufacturing Industries

According to Saraj Zadeh Fard (2016) studies, the highest prevalence rates of musculoskeletal symptoms were observed in health- care providing industry followed by manufacturing industries [22]. As shown in figure 1, the occupational diseases & poisoning by sector in Malaysia.

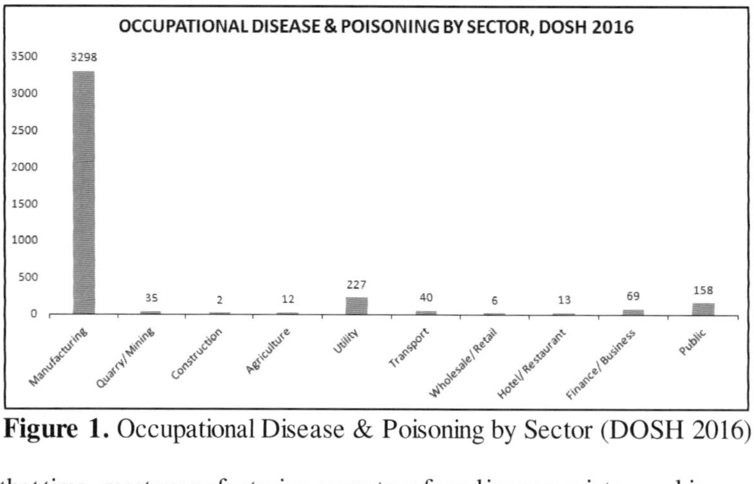

Figure 1. Occupational Disease & Poisoning by Sector (DOSH 2016)

During that time, most manufacturing operators faced inappropriate working conditions and this made it easy for operators to engage in panic-induced injuries or accidents, hot, stuffy and dusty environments, lack of ventilation and overtime work. [8]. Work-related MSD involves inflammatory and degenerative processes which can involve muscles, tendons, cartilages and joints, causing pain and impairment of function. [23]. Musculoskeletal symptoms risk factors are work activities such as awkward and static postures, handling materials manually, repetitive movements as well as individual, psychosocial, and organizational factors which known as important predictive variables at manufacturing industries lead to various types of symptoms [22] [24]. Figure 2 and 3 shows the some examples of the musculoskeletal symptoms among production operators at manufacturing industries.

Figure 2. Lifting in pinch causes hand/wrist pain

Figure 3. Awkward postures that causes low back pain

3.1 Low Back Pain (LBP)

Low back pain (LBP) is one of the most common occupational health problems in modern industrialized societies and accounts for a large number of days of compensation and disability for workers [25][26][27]. The NIOSH report stated that the entire body's awkward working postures, lifting and vibration were risk factors for symptoms of low back [15]. Physical factors, heavy lifting and repetitive hand movement were closely related to low back pain reporting. Working with repetitive hand movements or working with awkward hand postures may involve light material handling that would repetitively load the back and may result in increased risk of back pain. [11]. Studies have shown that workers may feel exhausted for the prolonged standing position, seats have not been provided as the task requires standing and awkward posture most of the time and this could be another possible explanation for the higher rate of musculoskeletal symptoms on back pain [4].

3.2 Hand/Wrist Pain

Computer work, heavy physical work and vibration were stated on the NIOSH report as risk factors for the development of wrist / hand pain. Eatough et al. (2012) stated that operators with higher job control are able to have greater autonomy and therefore take more breaks, help their muscles and reduce the risk of developing musculoskeletal symptoms associated with work [28]. In another case study conducted in the textile manufacturing industries, it was stated that poor posture in the workstation such as incorrect table and chair heights and unadjustable equipment and the highly repetitive nature of the work are likely to contribute primarily to the high incidence of MSSs. [3]. The table height requires the operator to flex and extend the arm repeatedly, and to flex much of the load [3][15] on the wrist bearing. Using a force on an individual or object can overload our muscles and tendons [16]. Kolgiri (2016) also stated that the force may come from gripping, lifting, pushing or pulling and this pinch grip produces 3-5 times more force on the tendons in the wrist than a grip with the whole hand. With excessive force the muscles are contracting much harder than normal, this can lead to stress on the muscles and joints especially in wrist.

3.3. Shoulder Pain

For workers with shorter working hours, the shoulder pain was significantly higher because the workers who had frequent lifting had a high prevalence of pain in the shoulders [29]. Risk factors such as repetition of work, hard work, and awkward position, twisted and bent position, working with arms raised to shoulder level, and handling heavy weights can also lead to disorder of the upper limbs, particularly of the shoulders. Studies have shown that the prevalence of shoulder pain in the general population may be as high as 6%-11% below 50 years of age, increasing to 16%-25% in the elderly [8].

4. Conclusion

To sum up, there was a high prevalence of musculoskeletal symptoms among manufacturing operators with a high proportion of symptoms in the lower back, shoulders, knees, and neck. Furthermore, from the point of view of the type of activity, the prevalence rates of MSs in all body regions were higher among operators with dynamic jobs as compared to those of workers with static jobs. Risk factors such as heavy load lifting, repetitive movements, long-term standing position, and extra force exertion appear to be causes of high prevalence of musculoskeletal symptoms among workers with dynamic jobs. In order for operators in the manufacturing industry to handle their work efficiently, further studies should focus on appropriate working positions.

Acknowledgement
Author would like to thank Faculty of Mechanical Engineering and Manufacturing, UTHM and Smart Driving Research Centre (SDRC) for support and providing facilities to accomplish the study.

References

[1] Mohammadi G 2013 Musculoskeletal complaints among high school teachers," *J. Musculoskelet. Res.*, vol. **16**, no. 2, p. 1350010.
[2] Choobineh A and Behzadi M 2009 Musculoskeletal problems among workers of an Iranian zinc industry. vol. **15**, no. 4, pp. 419–424.
[3] Sealetsa O J and Thatcher A 2011 Ergonomics issues among sewing machine operators in the textile manufacturing industry in Botswana, *Work*, vol. **38**, no. 3, pp. 279–289.
[4] Foong M C, Adon M Y and Rafee B 2014 Prevalence of Musculoskeletal Symptoms Among Production Line Workers in a Printing, *Int. J. Public Heal. Clin. Sci.*, vol. **1**, no. 1, pp. 109–117.
[5] Silva J M N da, Silva L B da, and Gontijo L A 2017 Relationship between psychosocial factors and musculoskeletal disorders in footwear industry workers, *Production*, vol. **27**, no. 0, pp. 1–13.
[6] Shiri R, Karppinen J, Leino-Arjas P, Solovieva S and Viikari-Juntura E 2009 The association between obesity and low back pain: a meta-analysis, *Am. J. Epidemiol.*, vol. **171**, no. 2, pp. 135–154.
[7] Chandrasakaran A, Chee H L, Rampal K G and Tan G L E 2003 The Prevalence of Musculoskeletal Problems and Risk Factors among Women Assembly Workers in the Semiconductor Industry," *Med. J. `Malaysia*, vol. **58**, no. 5, pp. 657–666.
[8] D. A. W. M. Van Der Windt *et al* 2000 Occupational risk factors for shoulder pain : a systematic review," pp. 433–442.
[9] Chiasson M E, Imbeau D, Aubry K and Delisle A 2012 Comparing the results of eight methods used to evaluate risk factors associated with musculoskeletal disorders, *Int. J. Ind. Ergon.*, vol. **42**, no. 5, pp. 478–488.
[10] Section P 2017 Latest OSH Statistics and SOCSO Programs.
[11] Waters T R, Dick R B, Davis-Barkley J, and Krieg E F 2007 A Cross-Sectional Study of Risk Factors for Musculoskeletal Symptoms in the Workplace Using Data From the General Social Survey (GSS)," *J. Occup. Environ. Med.*, vol. **49**, no. 2, pp. 172–184.
[12] Jaffar N, Abdul-Tharim A H, Mohd-Kamar I F, and Lop N S 2011 A literature review of ergonomics risk factors in construction industry," *Procedia Eng.*, vol. **20**, pp. 89–97.
[13] Kee D, Chung M K and Kim J H 2011 Legal system and its effect for prevention of work-related musculoskeletal disorders in Korea," *Int. J. Ind. Ergon.*, vol. **41**, no. 3, pp. 224–232.
[14] O'Sullivan K, O'Sullivan P, O'Sullivan L and Dankaerts W 2012 What do physiotherapists consider to be the best sitting spinal posture?, *Man. Ther.*, vol. 17, no. 5, pp. 432–437.
[15] Costa B R and Vieira E R 2010 Risk factors for work-related musculoskeletal disorders : A systematic review of recent longitudinal studies, *Am. J. Ind. Med.*, vol. 53, no. 3, pp. 285–323.

[16] Kolgiri S, Hiremath R and Bansode S 2016 Literature Review on Ergonomics Risk Aspects Association to the Power Loom Industry," *IOSR J. Mech. Civ. Eng. Ver. III*, vol. **13**, no. 1, pp. 2278–1684.

[17] Merisalu E, Mannaste M, Hiir K and Traumann A 2016 Predictors and prevalence of musculoskeletal disorders among sewing machine operators, *Agron. Res.*, vol. **14**, no. 4, pp. 1417–1426.

[18] Kaergaard A and Andersen J H 2000 Musculoskeletal disorders of the neck and shoulders in female sewing machine operators: Prevalence, incidence, and prognosis," *Occup. Environ. Med.*, vol. **57**, no. 8, pp. 528–534.

[19] Van L, Chaiear N, Sumananont C and Kannarath C 2016 Prevalence of musculoskeletal symptoms among garment workers in Kandal province, Cambodia," *J. Occup. Health*, vol. **58**, no. 1, pp. 107–117.

[20] Kittusamy N K and Buchholz B 2004 Whole-body vibration and postural stress among operators of construction equipment: A literature review," *J. Safety Res.*, vol. **35**, no. 3, pp. 255–261.

[21] Choobineh A, Daneshmandi H, Saraj Zadeh Fard S, and Tabatabaee S 2016 Prevalence of work-related musculoskeletal symptoms among Iranian workforce and job groups," *Int. J. Prev. Med.*, vol. **7**, no. 1, p. 130.

[22] Inbaraj L R *et al* 2013 Prevalence of musculoskeletal disorders among brick kiln workers in rural Southern India," *Indian J. Occup. Environ. Med.*, vol. **17**, no. 2, p. 71.

[23] Piranveyseh P *et al* 2016 Association between psychosocial, organizational and personal factors and prevalence of musculoskeletal disorders in office workers," *Int. J. Occup. Saf. Ergon.*, vol. **22**, no. 2, pp. 267–273.

[24] Hansson T and Jensen I 2004 Chapter 6. Sickness absence due to back and neck disorders," *Scand. J. Public Health*, vol. **32**, no. 63_suppl, pp. 109–151.

[25] Punnett L *et al* 2005 Estimating the global burden of low back pain attributable to combined occupational exposures," *Am. J. Ind. Med.*, vol. **48**, no. 6, pp. 459–469.

[26] Clays E, De Bacquer D, Leynen F, Kornitzer M, Kittel F and De Backer G 2007 The impact of psychosocial factors on low back pain: longitudinal results from the Belstress study," *Spine (Phila. Pa. 1976).*, vol. **32**, no. 2, pp. 262–268.

[27] Eatough E M, Way J D and Chang C H 2012 Understanding the link between psychosocial work stressors and work-related musculoskeletal complaints," *Appl. Ergon.*, vol. **43**, no. 3, pp. 554–563.

[28] Loo H S S and Richardson S 2012 Ergonomics Issues in Malaysia Centre of Ergonomics , Faculty of Mechanical Engineering , Faculty of Management , Multimedia University ," *J. Soc. Sci.*, vol. **8**, no. 1, pp. 61–65.

[29] Rehmana N and Charpeb N A, "Workplace Assessment and Ergonomic DRehmana, N., & Charpeb, N. A. (n.d.). Workplace Assessment and Ergonomic Designing of Checkstand.esigning of Checkstand."

IOP Conf. Series: Materials Science and Engineering **607** (2019) 012010 doi:10.1088/1757-899X/607/1/012010

Performance evaluation of wearable antenna printed using syringe-based deposition system on fabric

R K Khirotdin[1], M A Mahadzir[2], F C Seman[3], S H Dahlan[4] and N Hassan[5]
[1,2] Additive Manufacturing Research Group, Faculty of Mechanical and Manufacturing Engineering, University of Tun Hussein Onn Malaysia, 86400 Parit Raja, Batu Pahat, Johor, Malaysia
[3,4] Faculty of Electrical and Electronics Engineering, University of Tun Hussein Onn Malaysia, 86400 Parit Raja, Batu Pahat, Johor, Malaysia
[5] Faculty of Technology Engineering, University of Tun Hussein Onn Malaysia, 84600 Pagoh, Muar, Johor, Malaysia

E-mail: khairil@uthm.edu.my, adzeemmahadzir@gmail.com, fauziahs@uthm.edu.my, samsulh@uthm.edu.my, hafizzah@uthm.edu.my

Abstract. Wearable antenna is recently known as part of clothing for communication purposes. Syringe-based deposition system had been proven capable of printing functional antenna structure but the printed antenna performances have not been discussed in details especially on resisting bending under various angles, curvature radiuses and wet conditions. Hence, this paper discusses the performance of printed antenna under various bending angles, curvature radiuses and wet conditions. The effect of several bending angles, curvature radiuses and wet conditions on resonance frequency, return loss and radiation pattern were discussed and compared accordingly with simulation. The antennas were observed to resonate best at frequency of 1.571 GHz as expected and the return loss obtained was at -20 dB that is 30% more than simulation which offers better performance. The radiation pattern obtained was as predicted since larger magnitude for the main lobe was produced than the side lobe. The antenna exhibited exceptional elasticity as it resists bending up to 30° angle and 5 cm of vertical convex curvature radius. Polyester fabric performed as a good candidate for textile wearable applications and water had only minor effect on antenna performance makes the proposed antenna suitable for humid condition.

1. Introduction

Wearable electronic devices are becoming very popular in personal communications, computer systems and wireless monitoring of vital functions [1]. The main advantage is they are designed as elements of clothing capable of transmitting or receiving wireless signals [2]. Integrating antennas into clothing can improve the desirability of such systems, and novel methods of fabricating conducting structures on clothing will benefit of diverse applications [3]. This kind of incorporation will add the benefit of eliminating clumsy devices that can be entangled to some extent [4]. Syringe-based deposition system with conductive ink had proven capable of printing functional electronics structure previously via strain gauge sensor and antenna on fabric [5, 6, 7] as it provides flexibility [5] and

[1] khairil@uthm.edu.my

compatible with different viscosity of materials since current printing techniques suffered from ink incompatibility especially on conformal substrates due to its rigidity and low flexibility [8]. The antenna structure is intentionally printed on life jacket to constitute as part of a rapid response device during emergency. Thus, the study is to investigate the printing of silver conductive ink track that makes up the antenna structure on fabric using syringe-based deposition system and at the same the performance of the printed antenna structure were evaluated since they have not been discussed in details before especially on resisting bending under various angles, curvature radiuses, and wet condition. The technique deposits inks using filamentary concept and the ink is required to be cured using mainly heating process via oven.

2. Methodology

2.1 Specification of Antenna Design

A type of micro-strip patch antenna was first identified and designed. The detail dimensions were determined and modelled using computer-aided design (CAD) software as shown in figure 1 respectively. The antenna is designed to operate at resonance frequency of 1.575 GHz as required by Multimedia Communication and Malaysian Commission (MCMC) for rescue signal. There were four essential parameters specified including the operating frequency (f_o), dielectric constant of substrate (ε_r), the height of the dielectric substrate (h) and the thickness of the conductor patch (t) as shown in table 1. The total length of the antenna designed is 158 mm long with a 0.80 mm width of track.

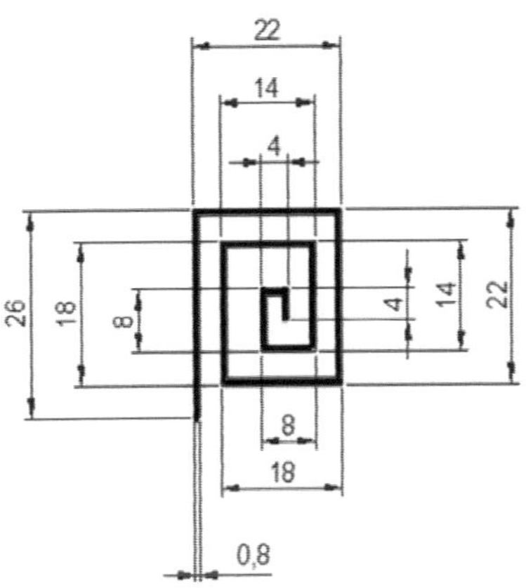

Figure 1. Micro-strip patch antenna designed with its detail dimension (units in mm)

Table 1. Specification of antenna designed.

Specification	**Width (mm)**	**Length (mm)**	**Thickness (mm)**
Dimension of ink track	0.8	158	0.22
Substrate (polyester fabric)	30	30	0.15
Ground plane	30	30	0.11

Table 2. Selection of printing and curing parameters

Printing Parameter				
Pressure (kPa)	Speed (mm/s)	Height Deposition (mm)	Nozzle Diameter (mm)	Viscosity (Pa.s)
160 - 240	5 - 9	1.20	0.51	2.40
Curing Parameter				
Temperature (°C)			Time (minute)	
160			45	

2.2 Printing and Curing Process

The antenna was initially simulated using Computer Simulation Technology (CST) Microwave Studio software to measure its resonance frequency, return loss (S_{11} parameter) and radiation pattern intended, directional with bigger main lobe and smaller side lobe as required. An automatic syringe-based deposition system (Model: FISNAR 3-axis, F4200N.1) equipped with a teach pendant and industry's leading dispensing software is used to deposit ink on polyester fabric as shown in figure 2. Conventionally, most of printing techniques only capable of printing on planar substrates and some of them including screen printing and inkjet printing suffered from clogging [8] and limited type of materials to be printed with. A series of pilot tests were conducted to determine the range of suitable printing parameters [7]. The selected printing parameters were varied accordingly as depicted in table 2 to obtain the desired dimension of antenna. The other two parameters including height deposition and nozzle diameter were set to be constant. A total of 25 samples were printed from the simplification method made using Taguchi method. Figure 3 shows the antenna that had been successfully printed using syringe-based deposition system and cured with oven to reduce the percentage of unwanted material in conductive ink since conductive ink is comes in liquid and required to be cured usually by heating process to reveal its metallic contents [6]. The curing process played an important role to provide a good conductivity of the ink tracks. Several pilot tests were performed to determine the optimum curing parameters that result in the lowest resistance. The optimum curing parameter selected is shown in table 2.

2.3 Material and Substrate

A conductive ink used throughout the whole study is silver epoxy-based ink (Model: AG806) as shown in figure 4 and direct usage of the ink is not possible due to high solid fraction thus an adjustment of viscosity via dilution with toluene solvent is necessary. It is important to ensure that during filling the ink into the barrel, it is properly packed to prevent any air pocket left in the nozzle. The fabric material used is made of polyester since it is suitable to mimic the actual fabric made for life jacket

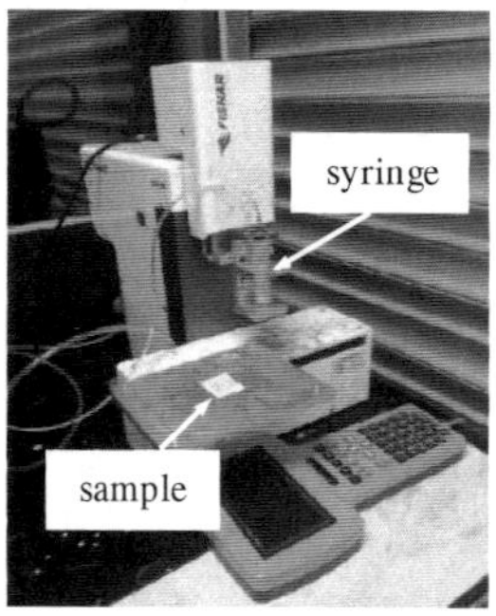

Figure 2. Automated syringe-based deposition system

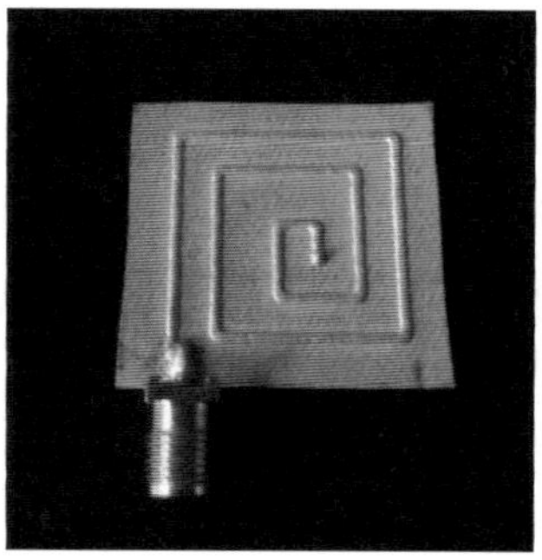

Figure 3. Successful micro-strip patch antenna printed on polyester fabric

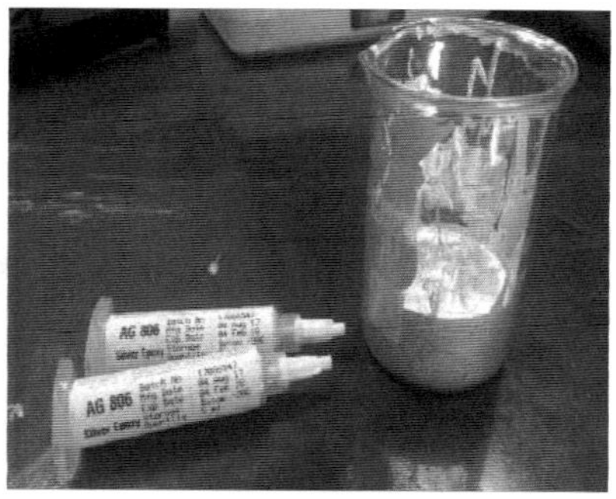

Figure 4. Silver conductive ink (Model: AG806)

2.4 Measurement.

Analysis of the antenna printed is consisted of measurement of width and thickness of the ink track using optical microscope. Its electrical properties were measured via inductance, capacitance and resistance (LCR) meter. A cross-sectional area of the ink track and its total length were determined accordingly to calculate the resistivity [6]. A measurement of resonance frequency and return loss (S_{11}) were determined using vector network analyser (VNA). The samples were tested next in semi-anechoic chamber to measure its far-field radiation pattern [7]. The antenna was aligned first to a horn antenna with adjustable polarization respectively. Since the antenna in a realistic on-body wearable antenna application will be subjected to bending and wetness thus, the effect of bending and wetness were investigated. All samples were tested on a customized curvature and bending jig as shown in Fig. 5 (a) and (b). The performances were tested on several curvature radiuses at two different positions; horizontal convex and vertical convex with different radius each; 2 cm, 3 cm, 4 cm and 5 cm. The bending performance were measured by applying bending angles from 0° to 180° and the test stopped when the ink track cracked. The samples were then measured in four different states of wetness involving fully wet, immediately wet, approximately dried and fully dried conditions. At first, the sample was kept under water for more than three hours and measurement was taken inside the water.

After that, the sample under test was taken out from the water and measured immediately. The sample was dried and measured again. Finally, the performance of the sample was tested when it is in fully dried condition. These four wet tests were illustrated in figure 7 (a), (b), (c) and (d).

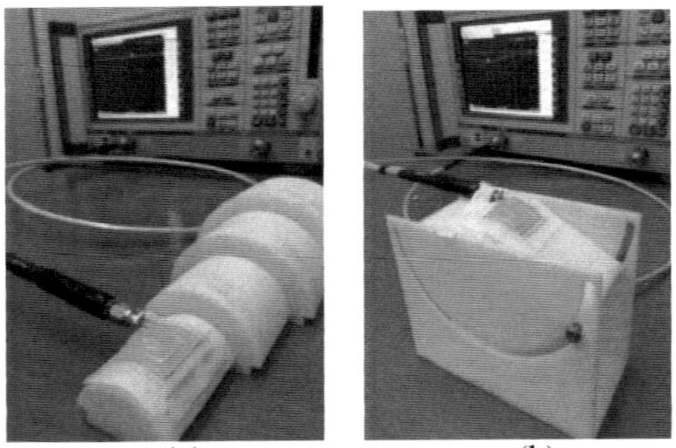

(a) (b)

Figure 5. Bending performance test on customized jig (a) curvature and (b) angle jig

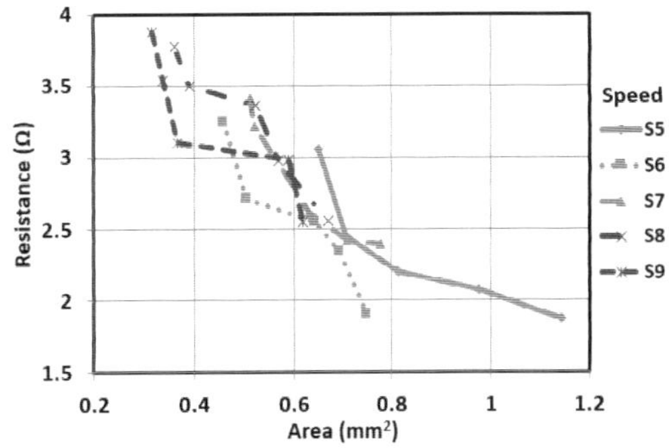

Figure 6. Relationship between cross-sectional area of the ink track with respect to resistance

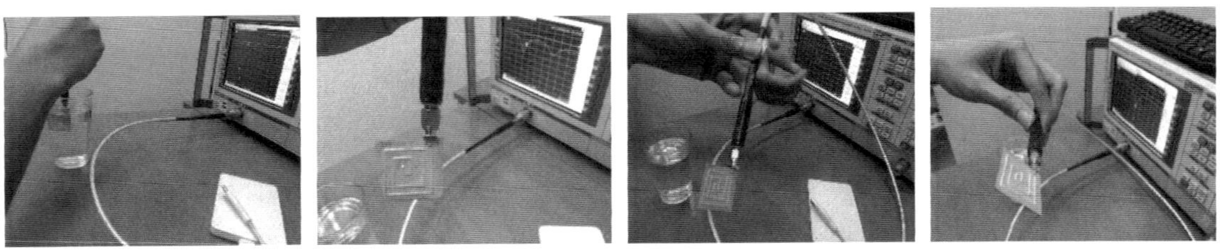

(a) (b) (c) (d)

Figure 7. Wet performance test conditions (a) fully wet (b) immediately wet (c) approximately dry (d) completely dry

3. Results and Discussion

3.1 Electrical Properties.

Figure 7 shows the relationship between cross-sectional areas of the ink track with resistance obtained. The trend shows resistance is decreased as the cross-sectional areas of the ink track increased. The relationship obtained also validated the general formula of resistivity [6] where bigger cross-sectional area provides lower resistance since bigger area allows more electron to flow through. Resistance is defined as the interruption of the flow of the electrical charge in the track. For an electron to travel from terminal to terminal is not a direct route but rather in a zigzag path due to countless collisions with fixed atoms within the conducting material. Thus, cross-sectional area of the ink track affected the amount of resistance. Wider and thicker ink track provides larger cross-sectional area thus less resistance for the electric charge to flow. Moreover, the total length of the ink track also affected the amount of resistance. The longer the ink track causes the resistance to be higher because the electron travelled in longer path and more collisions occurred.

3.2 Resonance frequency, Return loss (S_{11}) and Radiation Pattern.

The result of simulated and measured performance of the antenna on resonance frequency and return loss of one of the good samples was shown in figure 8. The simulated antenna was radiated best at 1.575 GHz as required indicated by red line. The bandwidth expanded from the lower frequency of 1.54GHz to the upper frequency of 1.63GHz and the magnitude of the antenna was recorded at -14 dB. The measured antenna was observed radiated best at 1.571 GHz with a magnitude of -20 dB indicated by blue line. The antenna averagely radiates best at 1.5833 GHz with the average magnitude of -23.2 dB. Although there is a slight difference bandwidth of the frequency by only 2.87% from the intended frequency, but it is still in the range of the intended frequency. The percentage difference between the simulated and measured resonance frequency is by merely 0.254% which is considered low thus the antenna designed is comparable to be used in wearable antenna application. In addition, the magnitude of S_{11} parameter obtained from measurement indicated better performance as the magnitude is higher than the simulated by 30%. In the meantime, the simulated and measured radiation pattern of the antenna designed is depicted in figure 9 where the far-field directivity of the antenna by simulation was observed to be 3.77 dB at 1.575 GHz with the main lobe direction at 5.0 degree angle. Furthermore, the side level of the antenna is -2 dB. Besides, the measured result has a directional average directivity at 1.6203 GHz. The far-field directivity is recorded at 8.479 dB magnitude. The main lobe average angle is 6.66 degree. Both results proved fulfilled the requirement for the antenna designed previously.

3.3 Bending and Wet Performance.

The sample was tested on several bending angles and curvature radiuses and the results from table 3 shows that the sample was still resonated but not at its designated resonance frequency starting from 10° to 30° angles. But the sample was not resonated at any frequency after it was bended at 40° angle and more because of the ink track was cracked during experiment starting from 40°. It was observed that when the sample was bended, the resistance changed and the amount of changes is proportional to how big is the angle of bending and this cause the resonance frequency to shift. Similar case happened when the sample was bended by curvature radius when the only sample that resonated at frequency of 1.622 GHz was when it was less curvature bended at vertical convex position with radius of 5 cm. The measured return loss of S_{11} parameter acquired was at 22.429 dB. In the meantime, table 4 shows the results for the wet performance test. The antenna did not resonate and function well at any frequency when the antenna was placed under water (fully wet) for 3 hours while in immediate wet condition, the antenna resonated but with slightly low frequency at 1.321 GHz and the measured return loss of S_{11} parameter was obtained at –11.554 dB respectively. Besides, the antenna was well functioned when in approximately dry condition which the antenna started to resonate back at its designated resonance frequency. But there are some negligible effects of water and moisture on antenna performance. The antenna resonated at 1.455 GHz and the measured return loss of S_{11} parameter is at –19.709 dB. Finally, the performance of the antenna was tested under fully dried condition and the performance recorded is remained unchanged similar to the approximate dried condition.

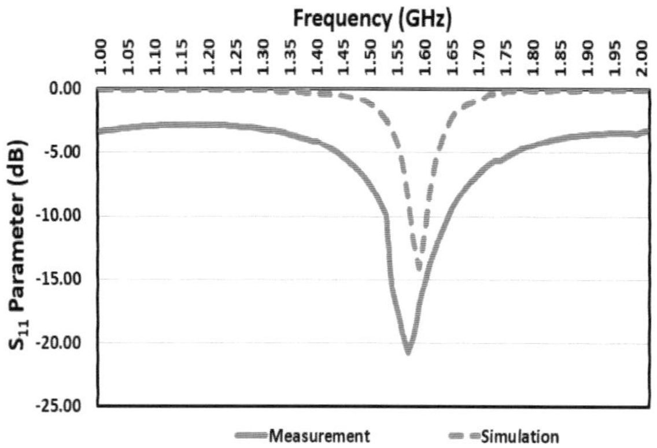

Figure 8. Result of simulated and measured of the antenna with respect to its resonance frequency and return loss, S_{11} parameter

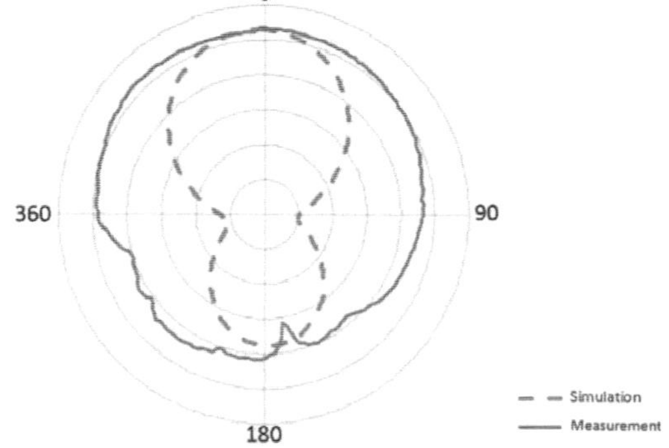

Figure 9. Simulated and measurement of radiation pattern

Table 3. Result of bending with curvature radiuses and angles.

Bending test	Angle, Position and Radius	Result	
		Frequency (GHz)	S_{11} parameter (dB)
Angle	0	1.586	-23.654
	10	1.628	-22.439
	20	1.662	-17..406
	30	1.662	-17..406
	40 -180	Not available	Not available
Curvature	Horizontal convex (radius = 2 – 5 cm)	Not available	Not available
	Vertical convex (radius = 2 – 4 cm)	Not available	Not available
	Vertical convex (radius = 5 cm)	1.622	-18.601

Table 4. Result of water performance test

Condition	Result	
	Frequency (GHz)	S_{11} parameter (dB)
Fully wet	Not available	Not available
Immediately wet	1.321	-11.554
Approximately dry	1.455	-19.709
Completely dry	1.455	-19.709

4. Conclusion

In conclusion, the antenna printed had proven could operate near to its designated resonance frequency. It could also be perceived that many aspects caused the measurement to have a slight difference by only a mere 0.254% from the simulation thus makes it compatible to be used for wearable antenna application. This slight difference may relate to fabrication tolerance and misalignment during measurement setup including imperfection in soldering of SMA connector which result in mismatching condition on port during testing. In addition, a fabrication tolerance also needs to be considered where the actual dimensional shape of the fabricated antenna is not the same shape with the designed of the ink track in simulation. However, the measurement offers a good performance since the return loss of S_{11} parameter and the magnitude of the main lobe of radiation pattern is higher than the simulation. Besides, it had proven that the syringe-based deposition system is capable of printing functional electronics structure particularly antenna since the wearable antenna printed shown comparable results. The functional wearable antenna operated and radiated well at its designated frequency. Moreover, effects of bending and wetness on antenna performances indicated that the used of conductive ink provides some elasticity hence makes the antenna bendable but still well-functioned and polyester fabric performed as a good candidate for textile wearable applications. Minor effect of water on antenna performance makes the proposed antenna suitable for humid and robust condition consequently, evaluation of long-term behavior and durability may well be considered in future.

Acknowledgement
The authors gratefully acknowledged the support to the University of Tun Hussein Onn Malaysia for giving the opportunity to attend and present at this conference.

IOP Conf. Series: Materials Science and Engineering **607** (2019) 012010 doi:10.1088/1757-899X/607/1/012010

References

[1] Sankaralingam S and Gupta B 2010 Development of Textile Antennas for Body Wearable Applications and Investigations on Their Performance Under Bent Conditions *Progress In Electromagnetics Research B* **22** 53–71

[2] Adel H, Wansch R and Schmidt C 2003 Antennas for a body area network *Antennas and Propagation Society International Symposium. IEEE, ST-Antennas for a body area network* **1** 471

[3] Dhupkariya S, Singh V K and Shukla A 2014 A Review of Textile Materials for Wearable Antenna *JoMET* **1** 7–14

[4] Ha S, Lee S, Yoo H J and Jung C 2010 Compact textile patch antenna for wearable fabric applications *Asia-Pacific Microw. Conf.* 2256–2259

[5] Khirotdin K, Ngadiron M F, Mahadzir M A and Hassan N 2017 Performance Evaluation of Strain Gauge Printed Using Automatic Fluid Dispensing System on Conformal Substrates *Material Science & Engineering Conference Series* **226** 1 012019

[6] Khirotdin R K, Hassan N, Siang H H and Zawahid M H 2017 Printing and curing of conductive ink track on curvature substrate using fluid dispensing system and oven *Engineering Letters* **25** 3

[7] Khirotdin R K, Mahadzir M A, Nazli M M M N, Seman F C, Dahlan S H, Alam C K A C K and Hassan N 2018 Printing of Wearable Antenna on Textile *MATEC Web of Conference, Malaysian Technical Universities Conference on Engineering and Technology* **150** 04001

[8] Khirotdin R K, Zakaria D A and Mohadhir M 2017 Printing and curing silver conductive ink tracks from modified on-shelf inkjet printer on fabric *International Journal of Mechanicals & Mechatronics Engineering IJMME-IJENS* **16** 6 50-56

IOP Conf. Series: Materials Science and Engineering **607** (2019) 012011 doi:10.1088/1757-899X/607/1/012011

Portable Mini Turntable for Close-Range Photogrammetry: A Preliminary Study

Adrian Savari Thomas[1,*], Mohd Fahrul Hassan[1], Al Emran Ismail[1], Reazul Haq Abdul Haq[1], Ahmad Mubarak Tajul Arifin[1], Md Fauzi Ahmad[2]

[1]Faculty of Mechanical and Manufacturing Engineering, Universiti Tun Hussein Onn Malaysia, 86400 Parit Raja, Batu Pahat, Johor, Malaysia
[2]Faculty of Technology Management, Universiti Tun Hussein Onn Malaysia, 86400 Parit Raja, Batu Pahat, Johor, Malaysia

*Corresponding author email: adrianthomas2806@gmail.com

Abstract. Photogrammetry is a science formed of technology exhibiting over a century of historical features and evolution. Moving forward on techniques employed in acquiring data from object portrayed in imaging has refashioned significantly from a pure optic-mechanical literature to an utterly digitize workflow nowadays. Parallel to this, the handling becomes elementary and thus it is practicable for non-photogrammetric naturalist to realize on this practice that the ability to develop 3D modelling from captured 2D photographic images is a noteworthy feature. However, capturing perfect photographic images comprising of all detailed textures and features is a troublesome task. Therefore, in acquiring higher quality and precision modelling, photography procedure techniques are vital to be comprehended. On the other side, an innovative portable mini turntable is planned to be designed as one of the shooting strategies to aid in the photogrammetry process. In this paper, a preliminary study on portable mini turntable development that purposely for close-range photogrammetry is discussed. This study is a part of the design process where several important considerations are identified to ensure the successful of the portable mini turntable development. Based on this study, the future works will be easily executed with a correct justification.

1. Introduction

Photogrammetry is an engineering code of practice that has spread by the impact of revolution in computer science [1]. Therefore, the drastic increment in computer utilization has heavily swayed on the photogrammetry application. This can be perceived through a perceptible shift from analog to logical reasoning and digital techniques [2]. The word 'Photogrammetry' is derived from three Greek words, whereby *phot*, *gramma* and *metrein* denote as light, drawing and measurement respectively [3, 4]. Primarily, photogrammetry is the science and technology of acquiring genuine information on surface texture and characteristic. The technique is engineered without subjecting any physical measurement via approaches such as recording, measuring and analysing from the captured photographic picture [5].

The propagation of photogrammetry techniques have become an apparent substitute in image-based modelling approach in consideration of the astonishing creation of pragmatic 3D models on top of its effectual course in communicating both the abstract and substantial idea or better-known as visualization [6]. Additionally, photogrammetry techniques are categorized into two groups of far-range, which branches into aerial and terrestrial techniques and close-range secondly, which this category was the main topic in this study.

1.1. Photogrammetry data acquisition, procedure and products

Essentially, there has always been an existing break that defines, as a technological gap between the recent discoveries in research on the other side and the execution of these outcomes in manufactured products that can predominantly lead in an industrial process. Concisely, photogrammetry has become a significant to technological term, whereby several organizations such as American Society for Photogrammetry and Remote Sensing, International Society for Photogrammetry and Remote Sensing and Committee for Documentation of Cultural Heritage (CIPA) are attached in this practice [4]. Notwithstanding on the resemblances between research and development, prominent dissimilarities are portrayed by which the results of photogrammetry studies are not publicly acknowledged yet.

Consequently, research institutes, research departments, universities, and certain mega industries are the only community that exploits the transition to these new emerging digital photogrammetry techniques. On the other side, countless of small and medium-scale industries are not willing to explore on these techniques yet due to self-censorship, micromanagement, inadequacy of expertise in the related field and also most of these such industries preferred on convectional techniques. Therefore, this accord to a time elapsed from the stage of an invention pioneering to a stage of reliable and workable photogrammetry application as illustrated in Figure 1 [2, 7].

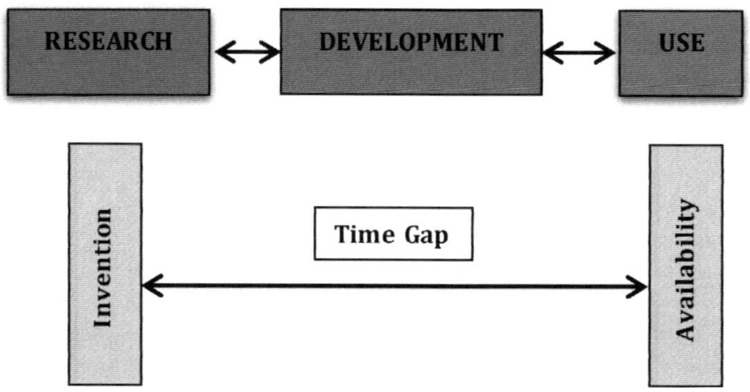

Figure 1. Time Elapsed from research pioneering to functional use of a new idea.

Digital close-range photogrammetry represents the principal means of acquiring 3D spatial database of an object from digital images. The utmost distinction between far-range and close-range photogrammetry is in its methodology, specialization and application. In far-range photogrammetry technique it encompasses an area surface (m^2) to be covered. Conversely, close-range technique emphasizes on physical object with its features dimensions, which varies between 1 cm^2 to 1 mm^2 [8]. By far, the abstraction of close-range photogrammetry in this paper is based on utilizing sequence-captured photos to computationally model 3D space. In sequence to comprehend on this subject, a disquisition on intertwined analysis portraying the complexity discipline of photogrammetry is presented, which is called the '*system approach*' as portrays in Table 1 [2]. Fundamentally, the input of this study is acquired via genuine information out of the process of recording electromagnetic radiant energy patterns, mainly of capturing photographic pictures [9]. After, this leads to an output, which is composed of a photogrammetric product outgrowing from several of procedures.

Table.1 Photogrammetry as in system approach.

Data acquisitions	Photogrammetric procedure	Photogrammetric product
	Softcopy workstation.	
1. Physical information. 2. Geometric information. 3. Semantic information. **4.** Temporal information	Dense image matching. (Semi-Global Matching (SGM), Dynamic programming method) ↓ Meshing ↓ Optimizing ↓ Texturing	1. Photographic products. 2. Maps: I. Topographic maps. II. Special maps. 3. Computational results.

Also, the acquired input is predominantly denoted as physical information. Consequently, the aforementioned data acquisitions practice in contact with the photogrammetric procedure (*black box*) that ensues to the outputs or desirable photogrammetric products [1]. Additionally, this system approach of photogrammetry is further introduced and discussed in section 2.

Today, 3D computer graphics has commenced a tie-up with the photogrammetry techniques in manufacturing technology that assist in the modelling of 3D space acquisition via 2D photographs [10]. Consequently, advancing onto replicating 3D models, as the photogrammetric product via additive manufacturing platforms (3D printing technology) [11]. This medium has gained its popularity due to its practical and applicable method, whereby generated mesh model from photographs can be conceptualized into 3D model via software and desktop computing power. Moreover, the momentary lapse of fieldwork span is an astounding feature and this pre-eminent procedure is elucidated as '*image based modelling*' (IBM) [6, 12]. Therefore, in realizing on this matter several of the common shooting strategies have been a breakthrough. For instance, this includes a high-tech end technique of photo booth with hundreds of cameras, moving a camera around the object or moving the object manually, flying a camera on a drone or an UAV and employing a technique of an automated turntable [13].

1.2. State-of-the-art in creating 3D models

In initiating an eminent 3D model from photographs, numerous photographs of the object have to be captured for digitizing [8, 12]. Besides, the captured photos have to envelop every detail of the object prompting to a better percentage of overlapping photos between them [14]. Precisely, about 60% overlap between photographs is advisable to ensure a corrective relative arrangement for processing [15]. A smart phone camera with a minimum image resolution of 8 megapixels is an initial hypothesis in generating the resultant 3D model. Additionally, this study drew a comparison, whereby an image resolution of higher megapixels is featured in another phone model. Technically, higher resolution corresponds to more image detail resulting in precise 3D model.

Therefore, it is anticipated that a shooting strategy for digital close range photogrammetry techniques in a 3D model can be identified and summarized through an investigation from literature. The information in terms of the required parameters and working principle in effective shooting phase of photogrammetry techniques will be referred, as to design a portable mini turntable that will be used as an assistive device during photogrammetry process. Primarily, a mini turntable rotates an object and set increment and triggers a stationary camera [11]. The advantage of a turntable is adapting the object to shoot at all sides thus capturing any every aspect of the object in a full loop [15]. On the other hand, the camera has to be translated in three different heights and thus making other several loops to acquire higher resolution textures. Additionally, the conceptual design is planned as the camera translation from one point to another, which is to maintain a consistency of the overlapping images captured. Thus, photogrammetry is a measurement friendly, whereby with a high resolution of captured photographs it is likely to generate a 3D model through 3D software and a high level of computing performance. The method is presumed to be much more advantageous compared to the 3D scanning process, where the output is fast, high precision and quality and provide a potential cost solution [16, 17, 18].

2. Three dimensional evaluation and reconstruction

The utilization of photographic images in the remake of 3D model acquires various interdependent techniques, as these will be presented in the upcoming subsections. Additionally, this photogrammetry procedure or phenomenon based methodology is called image based modelling (IBM) or photometric modelling [6]. Continuing on the line, these elaborations are the *pre-processing* and *data processing* phases in regard to the fundamental stages of reverse engineering that will be presented in section 3.

2.1. Dense image matching

The fabrication of a precise and reasonably pragmatic textured 3D model is practicable. Fundamentally, a dense point cloud is generated, whereby it compels to accomplish a dense pairing between oriented photographic images [19]. On the other hand, referring to the previous statement, conventional techniques are not feasible comparatively. This is due to the dependent correlation of small-scale imagery windows accompanying a one-dimensional epi-polar line that inherit the maximum disparity extent [20]. Aside from that, these modes presumed a persistent variances, whereby displaying inaccuracy at discontinuance regions and tend to obscure the object peripheries [21]. Figure 2 delineates the typical case of 3D point recognition from two images [19, 21].

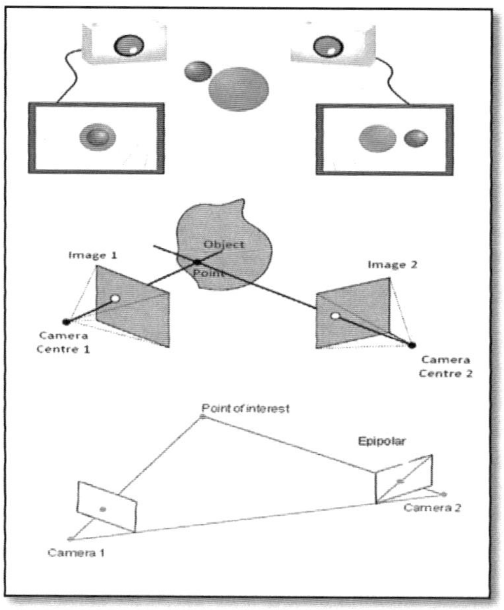

Figure 2. The postulation of 3D point recognition from two images.

Alternatively, advanced methodologies in realizing the eminent of dense image matching are being augmented. This progression focuses on acquiring a precise dense stereo matching particularly at object peripheries [21]. Also, emphasizing on the disclosing occlusions and resolving disparities via sub-pixel rendering are the other significant preference objectives. Essentially, there are two noteworthy advanced approaches that have been pioneered in the matching evolution as an optimization predicament, which are the application of Semi-Global Matching (SGM) and dynamic programming method (DP) [22]. However, the SGM technique is vastly advocated in many computational algorithms and modelling applications and by many researchers in the photogrammetric field considering from its beneficial functions to some degree [23].

The SGM technique plainly utilizes on the concept of pixel-wise matching of local and global stereo approaches at shorter runtime [23]. Thus, yielding a magnificent trade-off between precision and runtime peculiarly at the object boundaries [24]. Additionally, this auspicious technique is not affected by the variety of parameters and is robust against radiometric data. On that account, the SGM technique is ideally suited for most problem-solving methodologies and applications such as the 3D reconstruction [18]. In particular, point clouds are the paramount root data initiator for 3D acquisition and thus to acquire a dense point cloud, an analogous point is the prerequisite for practically every pixel in an image [14, 19].

2.2. Meshing and texturing
The final sequences in the progression of the 3D geometric model acquisition are meshing and texturing [25]. Primarily, miscellaneous methods are practicable in fashioning an absoluteness 3D mesh model commencing out of a dense point cloud. Hence, these methods can be categorized as in Figure 3.

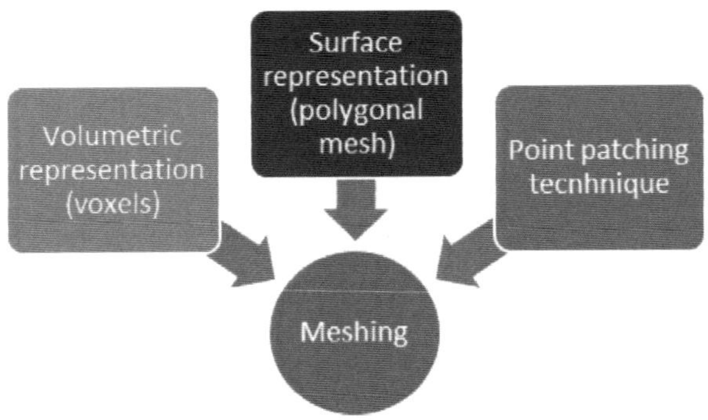

Figure 3. Meshing methods.

The volumetric representation by definition utilizes voxel in terms of unit volume factor. It was originated in the beginning of 1970's in the development of medical imaging and nowadays it is conventionally utilized in computer and scientific visualization and also computer graphics. However, the drawback on this point is the necessity for a tremendous storage volume in the meshing process [26]. On the other side, polygonal meshing is a compliant method to definitely portray the outcomes of 3D model valuation on condition that the surface description is adhered [27]. Likewise, point-patching technique is the construction of surfaces and curves directly from the points generated in the cloud [28]. It employs non-uniform rational basis spline (NURBS) mathematical model in fabricating and portraying these parameterized surfaces and curves [29]. After generating a 3D mesh model, texturing is the definitive procedure in which the software outlines digital colour image onto a 3D surface and thus portraying a photorealistic representation [27].

3. Extension of photogrammetry: reverse engineering

Reverse Engineering (RE) or better known, as the back engineering is an ageing practice that fashion or develop a digitized computer model from an existing object, whereby it is facilitated via a capable computer-aided engineering that succour in engineering analysis exercise [30]. This conceptualization has become a momentous emerging discipline, whereby being demonstrative on multitude of tasks and researches. Notwithstanding on conventional engineering that reshape the engineering abstraction and model into real framework, in reverse engineering patterns are revolutionize into engineering models and visualization [31]. Also, reverse engineering is a procedure of replicating an existing object in the absence of any aided drawings. Moreover, this approach defines its availability between systems that acquired interface [32].

Additionally, with the accelerating high-end computer system and computer software, it is plausible to digitally stitch multitudinous photographic images into 3D models virtual representation. Hence, reverse engineering disintegrates and analyses an object in details to disclose essential parameters or concepts, for object modification purposes and to replicate. Above all, reverse engineering is a far ranging in any engineering fields.

4. Photogrammetry with a smartphone

The photogrammetry modus operandi gazes at sequence of photographic images as well as reconstructing the camera placements, orientation and lens characteristics for a better images acquisition. Additionally, camera calibration is a major feature for a photogrammetric naturalist, as evaluation and precision is essential for a proper 3D modelling. Wherefore, in the works of better photographic images for prominent photogrammetry upshots via smartphones, several instructions are necessitated as follows.

4.1. Lighting environs

For expedient photogrammetry upshots, it is advisable to sidestep from an environment, wherein the light conceives a strong contrast shadows upon the object during photography session. The fact is that shadows are adversaries during the reconstruction period, as this phenomenon decentres the images points in which prompting towards a failed reconstruction model. Thus, essential instructions for proper shooting environments are described further.

a. Shooting interiorly

Experiment the object in a fully closed room with consummate dispersed lighting, which does not create contrast lighting effects or shadowy gestures. Also, abstained from setting flash mode on camera and do not shoot images in a room, whereby curtains are not hung on window panes, as light may penetrates into the room.

b. Shooting exteriorly

Do not experiment the object at midday under strong glazing sun. However, it is prudent to shoot images as early in the forenoon.

4.2. Assure that the object is placed at the centre of the camera frame (photo mode), which sates relatively about 70% of the examine object within the full-frame. The prior proportion is achieved via utilizing grid mode feature on camera as, an ideal manner for better acquisition purposes. Thus, the proportionately 70% is estimated to comprise all the nine square grids on the camera with the 5th grid being the most filled and the rest are filled marginally of the examine object.

4.3. The captured photographic images have to be persistently sharp thorough the photography session. Consequently, in archiving this promptness the lens's aperture of a smartphone is featured by modes termed as Auto Exposure (AE) and Auto Focus (AF) lock or known as *AE/AF Lock* in which it activates the tone by locking onto the impeccable focus and consistency exposure magnitudes in the photographic images. Thus, the exposure remains invariably balanced during the similar acquisition period. Additionally, AE/AF Lock ensures that the captured photographs are highlighted with every details displayed. Also, high dynamic range or known as *HDR* imaging is activated to aid better acquisition in generating higher dynamic range brightness imaging.

4.4. Do not activate the flash mode and plainly tap on the focus feature mode on various scopes within the scenes, as this is to avoid poor acquisition as well as the reconstruction from being blobby Therefore, it is advisable to feature the focus mode at the centre of the 5th grid to where the object is screened mostly as well as portraying balanced lightning intensity from every angle in sequence to acquire the best texture.

4.5. Supposing the object is retained statically and only the camera is moved around the object during the photographing session, thus, it is expedient to mount the smartphone on a tripod. Advantageously, the tripod stabilizes the phone and allowing less shakes during the capturing period, as blurry images can be hindered.

4.6. On the other side, the exposure scale-level on the camera can be calibrated via focus mode. Herein, a desirable level of exposure for absolute "tack sharp" photographic images is archived.

4.7. Do not edit, crop, mask and filter the captured photographic images beforehand advancing into reconstruction phases or utilizing filter smartphone camera frame during the photography session. The point is that these images may bring about failures during the reconstruction period aftermath.

5. Preliminary data acquisition and analysis

For preliminary data acquisitions, the most common and accessible shooting strategy is moving the camera around the object. Additionally, for high quality results when shooting objects, the basis idea is to go around the object shooting full loops of images in increments of 5 to 15 degrees.

On the other side, if the object is complex and has lots of occlusions, it is advisable to shoot in smaller increments and to fill in any heavily occluded areas with extra images at the end. Essentially, any surface on the object must be visible in at least two or more photos to be reconstructed in 3D. Besides, fill in as much of the frame possible with the object during the photography session. Furthermore, for larger objects or to acquire higher resolution textures at the end, it is prominent to shoot closer to the object in overlapping rows. Additionally, two phone models with distinct camera lens resolutions of 8 megapixels and 12 megapixels are utilized respectively during the photographing session. Later, these images of dissimilar resolution are then worked with Autodesk Recap for a preliminary 3D reconstruction and the acquired results are portrayed in Figure 4 demonstrating the dissimilarities respectively. In this assay, a clutch master cylinder is utilized, as a model for a preliminary test.

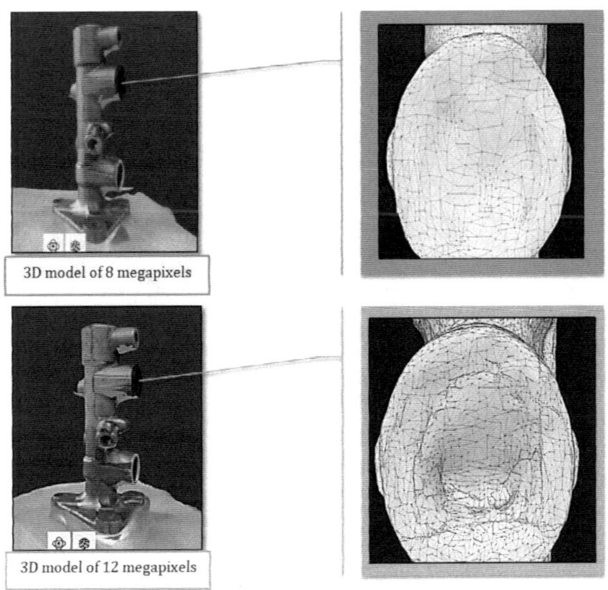

Figure 4. Preliminary 3D reconstruction results and analysis.

In Recap, the meshes properties for both 3D models taken from different images resolution are observed. The object shot by a 12 megapixels lens camera has higher number of mesh polygons that the visualizations is sharper compared to when it is shot by the 8 megapixels lens camera. Thus, this preliminary information describes that the higher a camera resolution the higher the mesh density. However, higher resolution images utilize a longer processing time and in future it may require much more RAM for smooth processing [33].

The contrast portrayed is based upon the photo resolution produced from the both smartphone models, as image resolution is determined by the capabilities of its digital camera. However, the results acquired are fairly satisfied, as the model is processed only with 80 imagines. Thus, the sequential photo redundancy will be enhanced in the upcoming 3D reconstruction through a portable mini turntable approach for a better acquisition [34].

6. Furtherance in developing a portable mini turntable

Development of the portable mini turntable can be accomplished through a heedful implementation of an engineering design process, whereby every aspect of characterization as well as specifications are inspected in detail before testing and verification of the product is made. Additionally, the photogrammetry turntable will be tagged, as a Pre-Production Prototype (PPP) for practical purposes that aid in the photogrammetry methodology. Notwithstanding of its appearance, the product design will be developed through a SolidWorks 3D CAD solution that delineates the detail design phase, which aids throughout the photogrammetry shooting strategy. Also, the turntable enigmas are defined initially, whereby its purposes and planning are stressed as well as including benchmarking of several conventional photogrammetry turntables. After inferring on particular instruction of the turntable conceptual design, concept generation is made clear. Additionally, brainstorming is the primary supplement of this stage after functional decomposition. Thus, the progression in developing a photogrammetry turntable is accomplished in line with the proposed workflow as depicted in Figure 5.

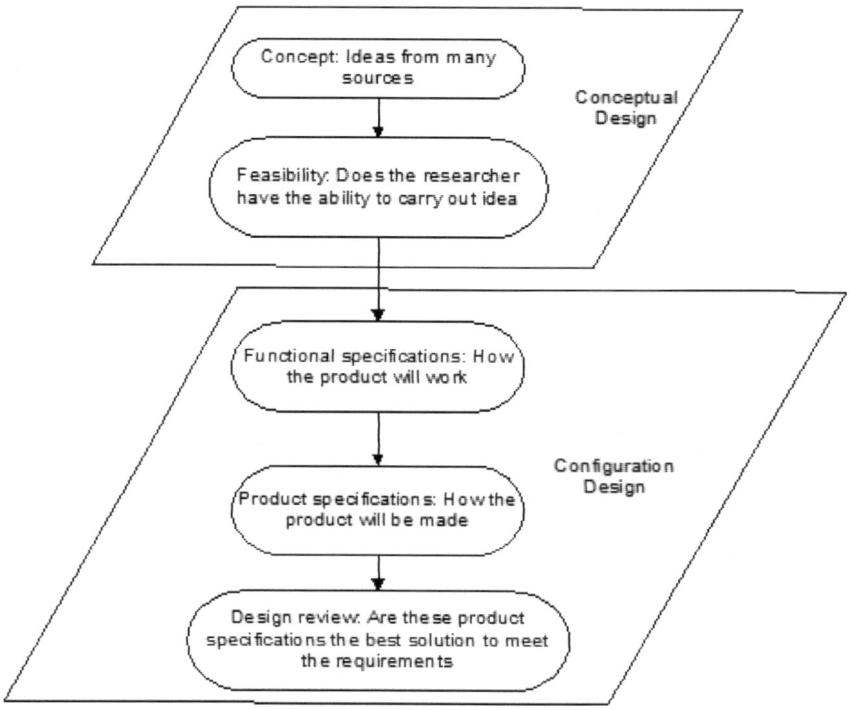

Figure 5. The proposed workflow of the fabrication procedure.

7. Summary and future works

The new-fangled photogrammetry method transfigures two-dimensional photographic images into three-dimensional spatial model through CAD software. At the same time, reverse engineering approach

IOP Conf. Series: Materials Science and Engineering **607** (2019) 012011 doi:10.1088/1757-899X/607/1/012011

is capacitated to acquire the digitization of substantial object inducting form the point cloud [35]. Subsequently, developing a digitized computer model from an existing object, whereby it is facilitated via a capable computer-aided engineering that succour in engineering analysis exercise. Nevertheless, this remarkable venture is often restrain due to the fact of limitations displayed in modern technologies. As the development of a turntable is succeeded, the major data acquisition is pursued with a turntable shooting strategy in accordance to the first objective of this research. A turntable rotates in a set of increments that triggers a stationary camera and the advantage of the turntable is shooting the object at all sides and thus capturing the entire object.

Without a doubt, this *Pre-Production Prototype (PPP)* turntable has to be developed through several potential design alternatives to ensure a suitable design is functionally efficient, easy to use and robust to response to inputs. As a result, embodiment design is the most vital phase in the fundamentals of engineering design study as this preliminary step provides a broad view of the existing design structures. Thus, Table 2 shows the functional factors and alternatives to be considered for the development of the portable mini turntable in this study. Therefore, these selections of criteria on the configuration design alternatives are based upon the consideration of design efficiency, sustainability and its authentic values in a serviceable environment.

Table.2 Functional factors requirement for the development of prototype turntable.

Product functional factors	Potential alternatives		
	Alternative 1	Alternative 2	Alternative 3
Gain energy	Batteries	Power Socket	USB cable
Energy conversion	Synchronous motor	Servo motor	Stepper motor
Case frame material	Wood	Plastic	Aluminium
Phone stand material	Wood	Acrylic tap plastics	Aluminium
Power transmission	Gear drive	Direct drive	Rubber belt drive
Turn ON/OFF switch	Toggle button	Push button	Switch
Enabling rotational motion	Square rotating swivel plate turntable bearing	Revolving swivel plate turntable bearing	Fidget spinner ball bearing
Assembly and disassembly	Rivets	Self tapping screw	Wood dowels
Type of microcontroller	ArduinoUno (Automatic rotation)	Raspberry Pi3 (Automatic rotation)	Manual rotation
Rotation speed	30 RPM	20 RPM	150 RPM
Angle of rotation increment	Rotates continously in full loops	0 to 5 degrees	15 degrees
Arm extension	Fixed at 50 mm	Moves between 75 to 155 mm	Moves between 250 to 1350 mm
Dimensions (WxDxH) of the turntable prototype	(412x474 204) mm	(235x235x82) mm	(600x600x200) mm

Overall turntable weight	3.0 kg	5.0 kg	10.0 kg
Max. load capacity (centered) for the turntable to rotate smootly	3.2 kg	2.0 kg	159.0 kg

Acknowledgments

The authors wish to thank the Research Management Centre, UTHM for the financial support to this project through the Contract Grant funding number H284 and the research fund to this work through the Postgraduate Research Grant (GPPS) funding number H319.

References

[1] Linder W 2003 *Digital Photogrammetry: Theory and Applications*, First. Springer.
[2] Schenk T, 2005 *Introduction to Photogrammetry*.
[3] Linder W 2006 Digital photogrammetry, *Control*, vol. **XXXIV**, p. 219.
[4] Walford A 2017 Photogrammetry," [Online]. Available: http://www.photogrammetry.com/.
[5] Stoian I 2014 The use of digital Photogrammetry and remote sensing technologies to graphic data bases in, pp. 114–126.
[6] Quan L 2010 *Image-Based Modeling*. Springer.
[7] Nguyen H M, Wünsche B, Delmas P, and Lutteroth C 2012 3D Models from the black box: Investigating the current state of image-based modeling," *20th Int. Conf. Cent. Eur. Comput. Graph. Vis. Comput. Vision, WSCG 2012*, no. **PART 2**, pp. 249–258.
[8] Kenarsari A E, Vitton S J and Beard J E 2017 Creating 3D models of tractor tire footprints using close-range digital photogrammetry," *J. Terramechanics*, vol. **74**, pp. 1–11.
[9] Scholten F and Wewel F, "Digital 3D-Data Acquisition with the High Resolution Stereo Camera-Airborne (Hrsc-a)," vol. **XXXIII**, pp. 901–908.
[10] Alsadik B S A 2014 *Guided close range photogrammetry for 3D modelling of cultural heritage sites*.
[11] Kaufman J, Rennie A E W and Clement M 2015 Single camera photogrammetry for reverse engineering and fabrication of ancient and modern artifacts," *Procedia CIRP*, vol. **36**, pp. 223–229.
[12] Remondino F, Menna F, Koutsoudis A, Chamzas C, and El-Hakim S 2013 Design and implement a reality-based 3D digitisation and modelling project," *Proc. Digit. 2013 - Fed. 19th Int'l VSMM, 10th Eurographics GCH, 2nd UNESCO Mem. World Conf. Plus Spec. Sess. fromCAA, Arqueol. 2.0 al.*, vol. **1**, pp. 137–144.
[13] Lievendag N 2017 3D Scan Expert, [Online]. Available: https://3dscanexpert.com.
[14] Remondino F and El-Hakim S 2010 Turning Images into 3-D Models," *IEEE Signal Process. Mag.*, pp. 55–64.
[15] AUTODESK, "AUTODESK REMAKE," *AUTODESK INC.*, 2016. [Online]. Available: https://remake.autodesk.com/about.
[16] Sketchfab, 2015 How to set up a successful photogrammetry project,.
[17] Baltsavias E P 1999 A comparison between photogrammetry and laser scanning," pp. 83–94.
[18] Kolecka N 2015 PHOTO-BASED 3D SCANNING VS. LASER SCANNING – COMPETITIVE DATA ACQUISITION METHODS FOR DIGITAL TERRAIN MODELLING OF STEEP," 2015.
[19] Kodde M 2016 Dense Image Matching," *GIM Magazine*, 2016. [Online]. Available: https://www.gim-international.com/content/article/dense-image-matching-2.

[20] Lhuillier M 2008 Robust dense matching using local and global geometric constraints," *Proc. 15th Int. Conf. Pattern Recognition. ICPR-2000*, vol. **1**, pp. 968–972.
[21] Hirschmüller H 2005 Accurate and Efficient Stereo Processing by Semi-Global Matching and Mutual Information," *Comput. Vis. Pattern Recognition, 2005. CVPR 2005. IEEE Comput. Soc. Conf.* vol **2** pp. 807–814.
[22] Hermann S and Klette R 2013 Iterative Semi-Global Matching for Robust Driver Assistance Systems, pp. 465–478.
[23] Hirschmüller H, Buder M and Ernst I 2012 Memory efficient semi-global matching," *ISPRS Ann. Photogramm. Remote Sens. Spat. Inf. Sci.*, vol. **I-3**, no. September, pp. 371–376.
[24] Rothermel M and Wenzel K, 2012 SURE - Photogrammetric Surface Reconstruction from Imagery," *Proc. LC3D Work,* pp. 1–21.
[25] Spann J R and Kaufman K S, 2000 Photogrammetry Using 3D Graphics and Projective Textures," *Int. Arch. Photogramm. Remote Sens. Spat. Inf. Sci.*, vol. **33**, pp. 2–9.
[26] Kuzu Y *et al.*, 2004 Volumetric Object Reconstruction.
[27] Remondino F and El-Hakim S, 2006 Image-based 3D modelling: A review," *Remote Sens. Photogramm. Soc.*, pp. 269–291.
[28] Yong W and Liang C 2010 A Real time NURBS Surface Interpolator for 5 axis Surface machining," no. 3.
[29] X. Du X, Huang J, and Zhu L M 2015 A complete S-shape feed rate scheduling approach for NURBS interpolator," *J. Comput. Des. Eng.*, vol. **2**, no. 4, pp. 206–217.
[30] Eilam E, 2005 *REVERSING: Secrets of Reverse Engineering*. John Wiley & Sons.
[31] Kumar A, Jain P K and Pathak P M 2013 Reverse Engineering in Product Manufacturing: An Overview," pp. 665–678.
[32] Singh N 2012 Review Article REVERSE ENGINEERING- A GENERAL REVIEW," pp. 1–5.
[33] Lievendag N 2017 AUTODESK REMAKE (RECAP PHOTO) PHOTOGRAMMETRY REVIEW.
[34] PHOTOMODELER 2017 Factors Affecting Accuracy in Photogrammetry.
[35] Xiong B, Oude Elberink S, and Vosselman G, 2014 Building modeling from noisy photogrammetric point clouds," *ISPRS Ann. Photogramm. Remote Sens. Spat. Inf. Sci.*, vol. **II-3**, no. September, pp. 197–204.

IOP Conf. Series: Materials Science and Engineering **607** (2019) 012012 doi:10.1088/1757-899X/607/1/012012

Preliminary Study of Kansei Engineering (KE) for Societal Performance in Sustainable Product Design

Mohd Fahrul Hassan[1*], Eka Ariefzakwan Annuar[1], Sia Chee Kiong[1], Mohd Azwir Azlan[1], Mohd Nasrull Abdol Rahman[1], Md Fauzi Ahmad[2]

[1]Faculty of Mechanical and Manufacturing Engineering, Universiti Tun Hussein Onn Malaysia, 86400 Parit Raja, Batu Pahat, Johor, Malaysia.
[2]Faculty of Technology Management, Universiti Tun Hussein Onn Malaysia, 86400 Batu Pahat, Johor, Malaysia.

*Corresponding author email: fahrul@uthm.edu.my

Abstract: Nowadays, product development indicates that a product's success in the market is determined by the inclusion of consumer's technical needs for product design. However, implicit feelings and impressions by the consumers is quite challenging because they cannot be externally assessed. Therefore, the makers aim to figure out the factors that contribute to the purchaser's fulfilments of their product. This paper introduces indicators that can be used to fulfil consumers' needs specifically to handle the implicit needs of consumers in sustainable product design based on Kansei Engineering (KE) approach. KE is a Japanese design method used to translate feelings into product parameters and was used to look at the car design features of a consumer's dream car. Preferences of four design features (safety, conveniences, car interaction, and size of capacity) were explored in a sample population of sixty university students through questionnaires. Seven kanseis/feelings elicited by phones were determined to be important to this group: (1) Dynamic, (2) Easy, (3) Elegant, (4) Personality, (5) Positive physical traits, (6) Sophisticated, and (7) Technology. This study chooses to explore the use of Descriptive Statistical and Correlation Component Analysis for statistical significance of the data set as the main method of analysis to determine whether there is a significant difference between the car design features and between the society's rating of the car features through Kansei words. Utilizing the system will help another automobile plan that matches consumers' needs to be proposed.

Keywords: Product design; Sustainability; Implicit needs; Kansei Engineering.

1. Introduction

Generally, product design plays a key role in the product development process for various purposes and can be approached in many different ways which evolve over time [1]. Evolutions in product design have led to many inventions, resulting in fairly good quality products flooding the market [2, 3]. Therefore, the competition is increasingly growing, forced to by the demanding market. Producers strive to design products that stand out and attract consumers. Currently, consumer's satisfaction and sustainable aspects (e.g., environment, economic and social) are equally important in determining the success of product design. However, product designer has thus far focused on social sustainability to a much lesser extent which is unfortunate, since not only can social sustainability practices help improve other aspects of sustainability, but all three dimensions are required to build a truly sustainable business [4, 5].

A significant amount of research has been generated in the last decades to support sustainable product development and help product design teams find engineering solutions to improve the environmental, economic, and social performance of products without compromising technical functionalities [6, 7]. Sustainable product design is about the implementation of sustainability considerations at the early stage of new product developments in order to produce a sustainable product. Baumann (2002) [8] identified more than 150 tools for 'green product development' that only focus on the environmental dimension of sustainability. Several researches have proposed sustainability frameworks that include all three dimensions, albeit with a greater emphasis on economic and environmental sustainability, and only a few have tried to examine social standards using empirical analysis [9].

Yang (2005) [10] identified product design as a strategic tool to be incorporated into sustainability solutions. Product design is responsible for ensuring that a design can produce a profitable product. Eco-design is a term for strategies that aim to integrate environmental aspects, and sustainable product design is more than eco-design, as it integrates the social aspect of the product's life cycle along with consideration of environmental and economic aspects [11]. This strategy makes product design an important element in the creation of the product for achieving sustainable development.

Overall, there has been limited investigation validating programs and investments to enhance organizational and supply chain social sustainability [12, 13]. Tools do exist to help with strategic and subjective evaluations of sustainability dimensions in organizations and their supply chains [14], but there is much to learn from different fields of study to strengthen the organizational strategic justification and forecasting toolset. The economics literature provides substantial groundwork and tools available for environmental valuations that may be applicable to organizational and supply chain social sustainability activities, both for internal organizational activities and across the broader supply chain players and actions. Therefore, the primary objective of this paper is to investigate the application of the Kansei Engineering (KE) method in discovering implicit consumers' needs that specifically applied in product design.

2. Methodology

This section comprises an in-depth discussion about the methods utilized in this exploration. This section also discusses the development of Kansei Engineering with several issues related to implicit consumers' needs that may influence sustainable product design. This method is described as a mental function, or more precisely, a higher function of the brain, and therefore it is implicit. For this study, KE Type II, KE Computer System is applied to compile databases and infer the main objective to support a computerized system that handles the process of interpreting consumers' feelings and emotions regarding perceptual design elements, which is described at the end of this section.

2.1 Kansei Engineering Approach

The suitable type of Kansei that matches the research objective is KE Type II which is the KE Computer System (KES). KES is a computer-assisted system. KES is an expert computerized system that can be utilized to translate the consumer's feeling and image into the design details. The KES architecture has four databases which include the Kansei database, containing the majority of the words related to consumer feelings which are illustrative of the product; image database which contains contributory things in the design details in the interest of a specific Kansei word, knowledge-based which contains the rules expected to choose items of the design details closely linked with the kansei words, and design and colour database which contains the design details with the colours separated [15].

So far, KE has been introduced to multiple industries including the automotive, construction machinery, electric home appliance, office machinery, house construction, costume, and cosmetic industry [15]. Application to mobile and entertainment devices have also been more recently explored [16, 17, 18]. These studies sought to find the emotional elements that are deemed significant to consumers. In addition to that, they were able to determine which factors influence consumer's impressions and which emotional tags were associated with each design element. If the model is found to be unsuccessful, the 'semantic' and 'product element' spaces should be updated until the model is able to yield reliable results as shown in Figure 1 which portrays a framework for the Kansei Engineering methodology. The basic idea is to describe the preferred features based on an earlier chosen domain which is the idea behind the product from two different perspectives, the semantic description and the description of product properties.

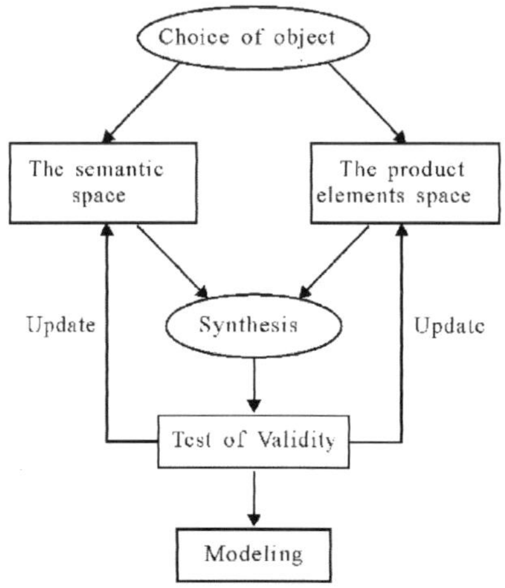

Figure 1. A Framework for Kansei Engineering [13].

Each of these two descriptions spans a vector space. Subsequently, these spaces are merged with each other in the synthesis phase, indicating which of the product properties evokes which semantic impact. Only after these steps have been carried out, that it is possible to conduct the validity tests, including several types of post-hoc analyses. As a result of this step, the two vector spaces have been updated, and the synthesis step is able to run again. When the results from this iteration process are satisfactory, a model can be built describing how the semantic and the space of application are associated.

3. Results and Discussion

This section shows the results of the data analysis and evaluation of the research according to the research objectives. Statistical calculations were used to test the research objective which was the KE's indicator for measuring the societal performance of product design using SPSS 20.00.

3.1 Application of Kansei Engineering for measuring societal performance of product design

In this study, an indicator was developed according to Kansei Engineering method. Figure 2 shows an indicator of Kansei Engineering for measuring the societal performance of product design. This indicator is created to facilitate the interpretation of a customer's feelings and readiness in design by using the KE method. With the KE method, there are various ways and procedures to obtain the output of the customers' desires. Therefore, with this indicator, the designer can easily analyse the feelings and requirements of the customers.

3.2 Case Study

The case study employed for this paper investigated a car as product design. There are seven car design features featured in this study after Questionnaire 1 was distributed. For Questionnaire 2, some design features had been explored to determine which ones will get the highest rating on the descriptive Kansei word.

3.3 Semantic Space analysis

From Questionnaire 1, 81 Kansei Words were collected from 60 participants. Furthermore, the KJ method was applied in this study to find a cluster of Kansei words by using an affinity diagram. The cluster name is based on the 10 Kansei affinity groups that have been selected from the previous study. In addition, 72 Kansei Words were collected from journals, magazines, and websites. Table 1 shows the cluster of Kansei Words after being combined from surveys and others (journals, magazines, and websites). From that table, selected cluster name uses in the analysis and synthesis case study is based on the major words collected in the cluster name. Based on Table 1, seven cluster adjectives were selected (Dynamic, Easy, Elegant, Personality, Positive Physical, Sophisticated and Technology) because they have a high number of adjectives compared to the other group names.

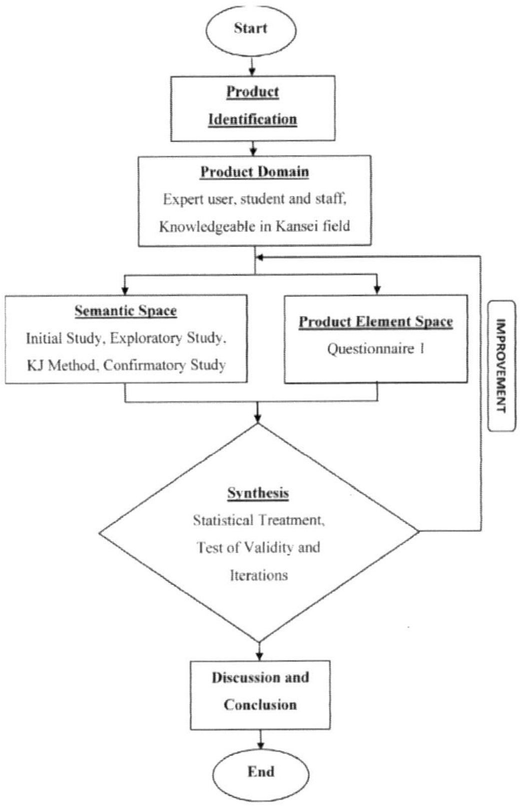

Figure 2. A Framework for Kansei Engineering [13].

Table 1. The Set of Kansei Words related to the car

Cluster Name	Kansei Words	Number of words
Adventurous	Capable, Energetic, Extreme, Valiant, Vigorous, Modern.	6
Basic	Common, Familiar, Plain, Simple, Standard, Typical	6
Creative	Innovative, Inspiring, Persuasive	3
Dynamic	Beneficial, Brand-new, Changeable, Flexible, Functional, High Potential, Latest, New, Talented, Useful, Strong, Aerodynamic	12
Easy	Easy-to-adapt, Easy-to-use, Just Nice, Manageable, Safe, Understandable, Easier	7
Elegant	Advance, Charming, Deluxe, Exotic, Exclusive, Gallant, Genuine, Gorgeous, Graceful, Impressive, Magnificent, Mature, Sexy, Smart, Sophisticated, Sporty, Stylish, Trendy, Well-known, Chic, Cute, Beautiful, Elegant, Valuable, Fantastic	25
Personality	Characteristic, Charismatic, Affordable, Low Cost, Compact, Cheaper, Premium, Compatibility, Small, Low Maintenance	10
Positive Physical	Firm, Fit for Gender, Good Colour Combination, Good Colour Tone, Good Condition, Good Design, Good Feeling, Good Looking, Good Pattern, Good Presentation, Good Quality, Good Shape, Good Style, Good Visibility, Neat, Tidy, Well, Well-Decorated, Well-Designed, Durable, Comfortable, Colourful, Ergonomic, Stainless	24
Sophisticated	Abstract, Branded, High Class, High Impression, Limited, Luxury, Stunning, Top-class, High Capacity, Future Design Technology	10
Technology	Fast, Geometric, High Performance, High-Power, High-tech, Multi-function, Secure, Specialized, Systematic, User-Friendly, Eco-friendly, Speedful, Efficient, High Speed, Economical	15

3.4 Product Element Space analysis
At this stage, the product components for this research were collected from journals, thesis, and websites related to the case study which ultimately is a dream car for the customers and consumers. The seven components of the product which are convenience, car interaction, car materials, the shape of the car, size of the car, capacity, and the safety features were assembled. Table 2 shows the source of relevant product components used in this study.

Table 2. The source of related product component

Product components on this research	Related product elements from the sources	Sources or references
Shape of car	Shape of Phone	Thesis/Kansei Engineering and cultural differences in mobile phone design by Loni M. Watson
Car materials	Phone materials	
Convenience	Navigation Tools	
Size of capacity	LCD Screen size	
Size of car	Size of phone	
Car interaction	Phone Interaction	
Safety	Driver killed in Tesla self-driving car	Headline News/ Driver killed in Tesla self-driving car crash ignored warnings, NTSB reports by Nathan Borney- Published 10:25a.m. ET June 20, 2017

From Questionnaire 1, the frequency of product components selected by the community is collected. There are 60 participants involved in selecting the best product components that they considered should be in their dream car. Only four design features were explored to determine which will obtain the highest ratings via the given descriptive Kansei words. This is because these four design features have higher frequencies than others, as illustrated in Table 3. Table 3 shows the list of product elements in the order of their frequencies.

Table 3. List of Product element

No.	Product Component	Frequency	Original Word List order
1	Convenience	35	Safety
2	Car Interaction	34	Convenience
3	Car Material	22	Car Interaction
4	Shape of Car	21	Size of Capacity
5	Size of Car	22	Car Material
6	Size of Capacity	25	Size of Car
7	Safety	37	Shape of Car

3.5 Synthesis Space
For the space mentioned above, the Semantic Differential Method is used in Questionnaire 2. In the following steps, the two spaces were merged, and a prediction model was built, connecting the semantic space and the space of product properties together. The predictive model produced the needs to be verified using various types of post-hoc tests. Subsequently, based on Questionnaire 2, the Satisfaction with Life Scale (SWLS) questions is analysed using the SPSS Statistic 20.00. The descriptive statistics method is shown in Table 4. The results show the mean and standard deviation of each SWLS question for 60 respondents.

3.6 Descriptive Statistics
Descriptive statistics are used in this study to describe a considerable amount of data that have been collected and translated into numeral values. This section discusses and analyses data that was gathered from SPSS 20.00 to determine descriptive statistics for scores of mean and standard deviation values.

Table 4 indicates the design features in terms of car safety. The findings indicate that the 'Easy' element with a mean value of 4.73 is the largest mean. The lowest mean value of 3.78 belongs to the 'Sophisticated' element. However, the highest SD value of 1.68 is obtained by the 'Sophisticated' element while the 'Easy' element has the lowest SD value of 1.22. For the car safety features, the 'Easy' element shows the highest result in terms of the mean, with a value of 4.73 and lowest standard deviation value of 1.22. The result also indicates that most of the respondents agree that the ease of operation is one of the important elements to look at before buying a car. The 'Sophisticated' element gains the lowest mean value of 3.78 and a standard deviation value of 1.68. This result shows that most of the respondents would not prefer a sophisticated look which is not beneficial to their car safety.

Table 4. Descriptive Statistics Result from SPSS Statistic 20.00

Satisfaction with a Life Scale		Mean	Std. Deviation
Safety	Dynamic	4.70	1.253
	Easy	4.73	1.219
	Elegant	3.88	1.519
	Personality	4.47	1.308
	Positive Physical	4.68	1.242
	Sophisticated	3.78	1.678
	Technology	4.37	1.377
Convenience	Dynamic	4.25	1.492
	Easy	4.18	1.546
	Elegant	4.17	1.317
	Personality	4.22	1.415
	Positive Physical	4.47	1.282
	Sophisticated	4.20	1.388
	Technology	4.42	1.430
Car Interaction	Dynamic	4.63	1.301
	Easy	4.67	1.284
	Elegant	4.67	1.217
	Personality	4.25	1.297
	Positive Physical	4.63	1.235
	Sophisticated	4.47	1.142
	Technology	4.48	1.200
Size of Capacity	Dynamic	4.42	1.381
	Easy	3.82	1.578
	Elegant	3.72	1.519
	Personality	3.77	1.544
	Positive Physical	4.12	1.508
	Sophisticated	3.75	1.385
	Technology	3.98	1.589
Total participant, N			60

3.7 Correlation Component Analysis

The Correlation Coefficient Analysis (CCA) is the measure of the strength of unity between the variables. In this study, the CCA is used to examine the similarities between the Kansei's responses to the community about the design elements of the car.

As evidenced by the simple correlation, the similarity of the Kansei responses regarding the design elements of the car are presented Table 5 shows the correlation of certain Kansei words in relation to the safety aspects of the design. For example, 'Dynamic' highly correlates with 'Easy', 'Elegant', 'Positive Physical' and 'Sophisticated', which means that the 'Dynamic' element in the safety of the car must be considered with the 'Easy', 'Elegant', 'Positive Physical' and 'Sophisticated' elements. The details are illustrated in Table 6, Table 7 and Table 8.

Table 5. Correlation Matrix for Safety

Correlation	Dynamic	Easy	Elegant	Personality	Positive Physical	Sophisticated	Technology
Dynamic	1.000	.657	.614	.480	.602	.487	.487
Easy	.657	1.000	.477	.536	.648	.402	.402
Elegant	.614	.477	1.000	.241	.304	.442	.442
Personality	.480	.536	.241	1.000	.792	.562	.562
Positive Physical	.602	.648	.304	.792	1.000	.773	.773
Sophisticated	.573	.319	.734	.340	.511	.702	.702
Technology	.487	.402	.442	.562	.773	1.000	1.000

Table 6. Correlation Matrix for Convenience

Correlation	Dynamic	Easy	Elegant	Personality	Positive Physical	Sophisticated	Technology
Dynamic	1.000	.657	.614	.480	.602	.573	.487
Easy	.657	1.000	.477	.536	.648	.319	.402
Elegant	.614	.477	1.000	.241	.304	.734	.442
Personality	.480	.536	.241	1.000	.792	.340	.562
Positive Physical	.602	.648	.304	.792	1.000	.511	.773
Sophisticated	.573	.319	.734	.340	.511	1.000	.702
Technology	.487	.402	.442	.562	.773	.702	1.000

Table 7. Correlation Matrix for Car Interaction

Correlation	Dynamic	Easy	Elegant	Personality	Positive Physical	Sophisticated	Technology
Dynamic	1.000	.676	.724	.517	.738	.687	.658
Easy	.676	1.000	.871	.733	.809	.755	.766
Elegant	.724	.871	1.000	.698	.809	.846	.762
Personality	.517	.733	.698	1.000	.714	.641	.629
Positive Physical	.738	.809	.809	.714	1.000	.833	.751
Sophisticated	.687	.755	.846	.641	.833	1.000	.822
Technology	.658	.766	.762	.629	.751	.822	1.000

Table 8. Correlation Matrix for Size of Capacity

Correlation	Dynamic	Easy	Elegant	Personality	Positive Physical	Sophisticated	Technology
Dynamic	1.000	.432	.413	.293	.749	.365	.366
Easy	.432	1.000	.918	.684	.536	.793	.688
Elegant	.413	.918	1.000	.701	.599	.827	.735
Personality	.293	.684	.701	1.000	.347	.772	.661
Positive Physical	.749	.536	.599	.347	1.000	.550	.475
Sophisticated	.365	.793	.827	.772	.550	1.000	.722
Technology	.366	.688	.735	.661	.475	.722	1.000

4. Sustainability result

To get the desired results, only the analysis method and case study were changed to achieve the study objective which is to improve the sustainable design of the product in societal performance. There are many aspects to consider when using "sustainability" in KE. Based on the case study in this research, the aspects considered are the safety of the car, the convenience of car, car interaction and the capacity. These aspects are the consumer demands for a new car model. If these aspects are considered, the sustainability of the car model will increase.

From a social sustainability perspective, if this car is successfully invented with the incorporation of certain design features (Safety, Convenience, Car Interaction, Capacity), the probability that the customers will be fully satisfied, and the number of car users will increase. Based on this study, the data collection from Questionnaire 1 shows that the percentage of the four design features mentioned above has a high rating. However, sustainability in economic and environmental terms will also increase as these aspects are considered.

5. Conclusion

When a designer starts to make new items, they have to investigate new shapes or components that speak to the consumer's preferences which will satisfy their requests as well as their subjective preferences. Regularly while planning, fashioners utilize their innovative plan to develop a new outline for automobile design in light of their expertise and motivation. Nevertheless, consumers have their distinct language while depicting the proposed outline of their ideas.

The paper proposes a system that joins the surmising principles and Kansei with the execution of KE. The procedure empowers the identification of the outlined ideas and trademarks that approximates certain requirements of buyers to the motivations of the designers. The system has an intent to explore a technique for KE in by establishing verifiable purchasers' needs and examining their relations to reasonable item outlines and to produce a method for estimating the societal execution of the item plan. By utilizing this system, another auto's plan that matches the shopper's needs can be designed while the greater part of the outline enables the purchasers to pick configurations exclusively delivered by a designer's aptitude and instincts.

The concept and framework presented in this paper are based on previous studies involving Kansei Engineering and emotional effects. Kansei Engineering is an important solution to induce human affective responses into new product design. Recent car designs concentrate on the societal performance of sustainable product design. However, the increasing consumer demands have led to the need for designers to introduce designs of cars that induce human emotions. Of course, there is always considerable risk in adopting Kansei Engineering. Other than the culture that is known to be different even in the same region, indigenous characteristic will also need to be taken into count. A wide range of features, categorized as performance, attraction, and essentials, have been decided to be applied to current product design. The result will then improve the sustainability of product design, with an increase in the development of a societal performance indicator by using KE approach. A handful of features are successfully determined throughout the process, which if applied to the product design, will significantly satisfy users and subsequently improve its sustainability.

Acknowledgments

The authors wish to thank the Universiti Tun Hussein Onn Malaysia (UTHM) for the financial support to this work.

Reference

[1] Hassan M F, Saman M Z M, Sharif S and Omar B 2016 Sustainability evaluation of alternative part configurations in product design: weighted decision matrix and artificial neural network approach. *Clean Techn Environ Policy* **18**, 63-79.

[2] Lokman A M 2010 Design & Emotion: The Kansei Engineering Methodology. *Malaysian Journal of Computing* **1(1)**, 1-11.

[3] Haq R H A, Wahab M S and Wahit M U 2014 Impact test and bioactivity properties of polycaprolactone (PCL) by addition of nano-montmorillonite (MMT) and hydroxyapatite (HA). *Applied Mechanics and Materials* **446-447**, 1129-1133.

[4] Seuring S, Müller M 2008 From a literature review to a conceptual framework for sustainable supply chain management. *Journal of Cleaner Production*, **16(15)**, 1699-1710.

[5] Hassan M F, Saman M Z M, Sharif S, Omar B 2013 Evaluation of sustainability performance of

product design element concepts using analytic hierarchy process. *Applied Mechanics and Materials* **315**, 799-808.

[6] McAloone T C and Pigosso D C 2017 Ecodesign to Sustainable Product/Service-Systems: A Journey Through Research Contributions over Recent Decades. In: Stark R, Seliger G & Bonvoisin J (eds) Sustainable Manufacturing. Sustainable Production, Life Cycle Engineering and Management, Springer, Cham pp. 99-111.

[7] Husin M M, Mustapa M S, Wahab M S, Arifin A M T, Masirin M I M and Jais F H 2017 Flexural properties untreated and treated kenaf fiber reinforced polypropylene composites. *AIP Conference Proceedings* 1846, 020008.

[8] Baumann H, Boons F and Bragd A 2002 Mapping the green product development field: engineering, policy and business perspectives. Journal of Cleaner Production **10(5)**, 409-425.

[9] Eizenberg E and Jabareen Y 2017 Social Sustainability: A New Conceptual Framework. Sustainability **9(68)**, 1-16.

[10] Yang M C 2005 A study of prototypes, design activity, and design outcome. Design Studies, **26(6)**, 649-669.

[11] Hassan M F, Saman M Z M, Mahmood S, Nor N H M, Rahman M N A 2017 Sustainability assessment methodology in product design: a review and directions for future research. *Jurnal Teknologi* **79(1)**, 63-70.

[12] Cho C H, Laine M, Roberts R W and Rodrigue M 2015 Organized hypocrisy, organizational façades, and sustainability reporting. Accounting, Organizations and Society **40**, 78-94.

[13] Marwah O M F, Shukri M S, Mohamad E J, Johar M A, Haq R H A and Khirotdin R K 2017 Direct Investment Casting For Pattern Developed By Desktop 3D Printer. *MATEC Web of Conferences* **135**, 00036.

[14] Bai C and Sarkis J 2014 Determining and applying sustainable supplier key performance indicators. Supply Chain Management: *An International Journal* **19(3)**, 275-291.

[15] Nagamachi M 2002 Kansei engineering as a powerful customer-oriented technology for product development. *Applied Ergonomics* 289-294.

[16] Lai H H, Lin Y C, Yeh C H and Wei C H 2006 User-oriented design for the optimal combination on product design. *International Journal of Production Ergonomics* 253-267.

[17] Chen K, Chiu S and Lin F 2007 Kansei design with cross cultural perspectives. In Aykin, N. (Ed.) Usability and internationalization: HCI and culture. U. Berlin; New York: Springer.

[18] Roy R, Goatman M and Khangura K 2009 User-centric design and Kansei engineering. CIRP *Journal of Manufacturing Science and Technology* 172–178.